Numanities - Arts and Humanities in Progress

Volume 1

Series editor

Dario Martinelli, Kaunas, Lithuania

The series originates from the need to create a more proactive platform in the form of monographs and edited volumes in thematic collections, to discuss the current crisis of the humanities and its possible solutions, in a spirit that should be both critical and self-critical.

"Numanities" (New Humanities) aim to unify the various approaches and potentials of the humanities in the context, dynamics and problems of current societies, and in the attempt to overcome the crisis.

The series is intended to target an academic audience interested in the following areas:

- Traditional fields of humanities whose research paths are focused on issues of current concern;
- New fields of humanities emerged to meet the demands of societal changes;
- Multi/Inter/Cross/Transdisciplinary dialogues between humanities and social and/or natural sciences;
- Humanities "in disguise", that is, those fields (currently belonging to other spheres), that remain rooted in a humanistic vision of the world;
- Forms of investigations and reflections, in which the humanities monitor and critically assess their scientific status and social condition;
- Forms of research animated by creative and innovative humanities-based approaches;
- Applied humanities.

More information about this series at http://www.springer.com/series/14105

Dario Martinelli

Arts and Humanities in Progress

A Manifesto of Numanities

Springer

Dario Martinelli
International Semiotics Institute
Kaunas University of Technology
Kaunas
Lithuania

Numanities - Arts and Humanities in Progress
ISBN 978-3-319-45552-5 ISBN 978-3-319-45553-2 (eBook)
DOI 10.1007/978-3-319-45553-2

Library of Congress Control Number: 2016950359

Printed on acid-free paper

This Springer imprint is published by Springer Nature
The registered company is Springer International Publishing AG
The registered company address is: Gewerbestrasse 11, 6330 Cham, Switzerland

Preface

> The humanities remind us where we have been and help us envision where we are going. Emphasizing critical perspective and imaginative response, the humanities—including the study of languages, literature, history, film, civics, philosophy, religion, and the arts—foster creativity, appreciation of our commonalities and our differences, and knowledge of all kinds. The social sciences reveal patterns in our lives, over time and in the present moment. Employing the observational and experimental methods of the natural sciences, the social sciences—including anthropology, economics, political science and government, sociology, and psychology—examine and predict behavioral and organizational processes. (Broadhead et al. 2013: 9)

A spectre is haunting Europe: the spectre of Numanities.

OK, not necessarily. But if you write a manifesto, you cannot help thinking about *that other* manifesto. Maybe that spectre *will* haunt Europe, at some point, but first there are many things that need to happen. To begin with, Numanities should become a spectre, in that particular sense that Marx and Engels meant.

More seriously, and more realistically, the present monograph was written to introduce a few proposals for a reformation of the humanities, in a way that comprises both a critical analysis to their current situation (by many called a "crisis"), and a full evaluation (rediscovery?) of their potentials. "Two things result from this fact" (as *that other manifesto* put it, too):

(1) This book presents the platform of Numanities in a systematic (though still embryonic) sense to the academic community. There have been journal articles, lectures, congress presentations, and in fact whole congresses (the International Congress of Numanities that takes place every year at Kaunas University of Technology), but a scientific monograph is still the most solid and effective way to launch a new theoretical program.
(2) This book kicks off the whole Springer book series in Numanities, with a first, "propaedeutic" I shall say, text that will hopefully set a bit of a template and trace a few patterns on the contents of the books that will follow.

The text, generally speaking, is divided into two main parts, plus accessories. The first part, corresponding to the second chapter, is a presentation of the actual program of Numanities, a manifesto in seven points, through which I attempt to lay the first tiles of what will hopefully be a long path, full of additions, refinements and corrections to these early reflections. The second part, corresponding to the third chapter, contains two case-studies which, drawing from my personal experience as a researcher (and I shall comment on the partiality of this choice), intend to show Numanities "in action", so to speak. Accessories include a "background" summary of the current crisis of the humanities, corresponding to the first chapter, and an appendix (which, strictly speaking, should have been another case-study) that illustrates the pre-history and the early history of Numanities, in relation to the main context it was generated within, i.e., the International Semiotics Institute.

Stylistically speaking, I wrote these pages following two specific genres of intellectual discourse: the pamphlet and, most of all, the manifesto. I think it is pretty important to specify this, in order to set the right expectations of the readers, and also—once again—to promote humanistic practices at their best.

On the one hand, indeed, the readers should know that the book assumes the typical programmatic and thought-provoking tones of these two (not so popular anymore) literary genres. This monograph is a manifesto because—let alone its title—it intends to offer a statement of purposes for this new forum called Numanities. Likewise, it is a pamphlet (albeit one that is too long for the usual standards) because it contains a lively tone that occasionally verge into the polemical one (here or there reminding us also of that particular, politically oriented, sub-genre called the libelle). I hope the readers will understand that such tones are, so to speak, part of the game, and that what I want is to put as much food for thought as I can on the table of this forum, in relation to my own competences, which are obviously limited, and far from being representative of the whole spectrum that—in my hopes—Numanities could and should reach. I will occasionally provide examples of research and scholars that —to my mind—are good opportunities to redefine the paradigm of humanities, and I will occasionally do it in a rather critical manner. However, I will not do it with the intention to "attack" them (and definitely not at personal level), but only as tools to make my point. Moreover, the people I will mention are already aware of my critical remarks, which have already emerged in previous congresses, publications and private conversations. The intention is to activate a chemical reaction (the debate on how the humanities could overcome their crisis) of the type that needs a high temperature (the polemical tone) in order to work out. So, I hope the readers keep this in mind while approaching this book (and do not think that I am this "angry young man" of the humanities—in fact, not that young either), and I hope that the colleagues mentioned forgive me for using one more time their cases as examples of my own arguments. I sincerely apologize in advance to these two groups if I do not succeed in conveying my benevolent intentions.

Another reason for engaging in these two genres lies in the fact that they both represent, in different periods of time, a form of intellectual practice where the humanities gave (and received) their best. My favorite example remains that of the manifesto of Futurism (the general one, that is—Futurism has produced an impressive amount of manifestos about nearly every topic). Let us forget for a second the (relatively true) allegations with Fascism and in general violence that this artistic movement produced and developed in time ("We will sing of great crowds excited by work, by pleasure, and by riot…"): I am interested in emphasizing the attention that surrounded its birth and early steps. Filippo Marinetti finalized this eccentric manifesto on February 5, 1909, and on the same day the document made the headlines of Bologna's *Gazzetta dell'Emilia* newspaper. During the next ten days the news spread all over Italy, and Napoli's *Il Pungolo*, Mantova's *Gazzetta di Mantova*, Verona's *Arena*, Trieste's *Il Piccolo* and Roma's *Il Giorno*, in this chronological order, covered the event. Finally, and most famously, on February 20 the manifesto was published in the main French newspaper *Le Figaro*. I used the word "event" to describe this media coverage, not just because Futurism later turned out to be one of the most important artistic movements of the whole century, but exactly because we are invited to stop and reflect on how central the arts and humanities must have been in early twentieth-century society, if the birth of an unknown artistic movement, promoted by a relatively unknown poet (Marinetti's first literary works were certainly more controversial than successful), managed to make the headlines of nonetheless than seven different newspapers, including some very important ones of national diffusion.

It is something I always mention to my students, whenever I have artists among them (and it happens pretty often, as the main subjects I teach are of the musicological and filmological type). "Can you imagine a similar situation nowadays?—I ask them—Let's say you decide today to start a new artistic movement—and I know that some of you are indeed involved in innovative forms of artistic expression. You sit down around a table, with a coffee or a tea, you establish a bunch of rules you commit yourselves to follow, and finally you write down a manifesto. Now, imagine you send this document to a national newspaper. Do you think they would even bother to reply to your email—let alone publish your document?" Students usually laugh at this point, and in that laughter there is the whole concentration of the crisis of the humanities in nowadays' society: a reaction so symptomatic of the loss of social relevance that artistic and humanistic practices experience today, that my students do not even consider anger or sadness as a response, but straightaway amusement. Paraphrasing that old joke about Berlin and Vienna during the Great War, humanities went from a situation that was serious but not desperate, to one that is now desperate but not serious.

However, to write a manifesto today is not only a nostalgic homage to the "glorious past" (also because artists and intellectuals still write them: the genre is less fashionable but far from dead). I rather consider it a nod to the fact that creativity, intellectual effort and "humanism" are not untied to a certain programmatic and pragmatic mentality. And, plus, they can also have the guts to speak with conviction, determination and clarity. Maybe, there is still some room for these qualities in the humanities, especially if we manage to complete them with a significant dose of humility, in order to avoid the risk that they develop into arrogance. By saying, as one does in a manifesto, "we intend to take the x, the y and the z action", we manage to achieve clarity (we do x, y and z, not quasi-x, meta-y and post-z), and at the same time to avoid arrogance (we will not pretend to be able to do also a, b and c).

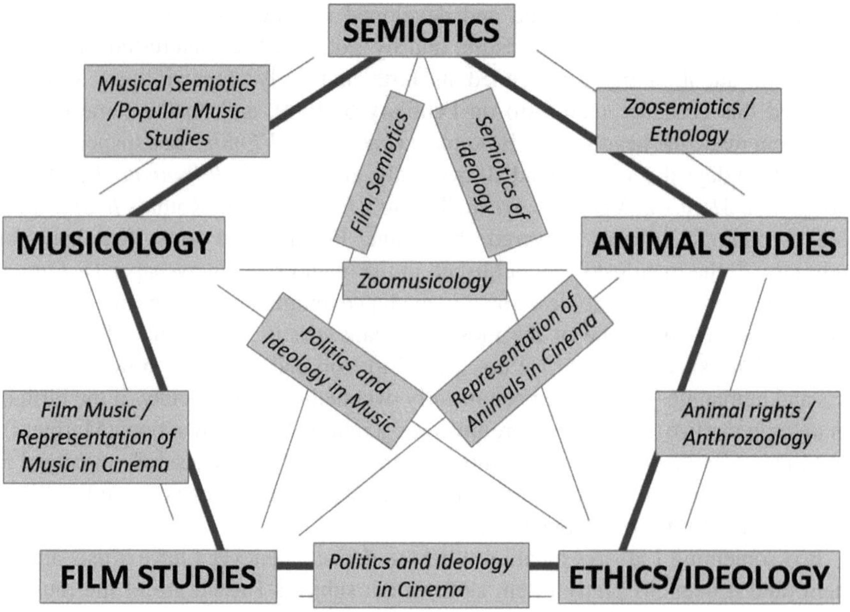

Fig. 1 Dario Martinelli's research map

As I have already hinted, a specific note should certainly be provided in relation to the methodological and the theoretical articulation of the two (or three, counting the appendix) case-studies proposed here. To say the least, the reader will find the latter rather partial and subjective, far from being representative of the whole spectrum that this book, on the contrary, seems to promise. What will soon be clear is that the vast majority of the case-studies are taken from fields like semiotics, musicology, animal studies and film studies.

The reason for this is quite simple: I did not want to present myself as a different scholar than who I really am. My research network is possibly a bit more varied than the average (not that this is necessarily an asset), yet at the same time extremely limited in comparison to the extensive size of the humanities as a whole. In Fig. 1 I even went as far as to map my fields of interest, in their position and interaction within my research path. I cannot really say if I have too many interests or too few: I have experienced both feelings in different situations. But one thing is for sure: whenever I remind myself of that old joke according to which there are two kinds of scholars, those who get so specific that they end up knowing everything about nothing, and those who get so general that they end up knowing nothing about everything, I know I lean more on the latter edge.

For any of the 15 fields and sub-fields represented in Fig. 1, I have taught courses and published essays (and in some cases monographs too). At a superficial level, one could say that my research path is very interdisciplinary, but the truth is, this applies only to a few of the cases mapped. I know I am being interdisciplinary when I deal with subjects like zoomusicology or zoosemiotics, and I definitely know I am not when I combine, say, musicology with film studies. Interdisciplinarity, as I will discuss later, is a word that everybody likes to attach to what they do nowadays, but it is becoming an abused concept, challenged by several forms of trivialization (as well as its related, but not similar, concepts of trans-, multi- and crossdisciplinarity).

That cinema and music can interact at a scholarly level is perfectly normal, and does not require a transfiguration of any of the two subjects, nor the flexibility to meet on a neutral field where a genuine symbiosis can occur: two cousins meeting for dinner are not exactly engaging in an intercultural experience, now, are they?

Back to my point, no matter how varied and occasionally interdisciplinary my personal research path can be, there is no doubt that, in the context of the present book, repeated mentions to areas such as semiotics, music, films or non-human animals may leave the reader with a few reservations on the (apparently limited) horizons of this whole enterprise. Even though that may be the case, it is important for me to mention that my approach to this problem has been of a different sort, and very much on the side of intellectual honesty. I did not want to speak about subjects I do not consider myself an expert in: I rather wanted to apply the paradigm of Numanities to what I do, hoping/expecting that whoever sympathizes with this program, will do the same with his/her own research path.

In fact, since we are touching the topic of my competences, it may be useful to emphasize the great "P-word" missing from my research map, that is, philosophy. There is very little philosophy in this book, all considered, especially if we intend this discipline as the practice of arguing one's cases by thoroughly going through the great philosophers of the past (particularly Hegel and Kant: for some reason, they are never missed). Of course, I like to think that philosophy is a wider form of reasoning than this, and that therefore there *are* significant philosophical arguments here. However, speaking strictly, this monograph has no classical philosophical approach. To make matters worse, the few philosophical arguments proposed here are never intended as a means to turn Numanities into a paradigm belonging to any particular philosophical school. When I mention passages of the likes of Marx or Dilthey, in the course of the text, it is not because I intend to make Numanities a "marxist" or a "diltheyan" theory: it is simply because those particular passages (and *not* the whole work of those philosophers) are particularly useful at that particular point.

Kaunas, Lithuania Dario Martinelli

Acknowledgments

To conclude, the inevitable, but very sincere, acknowledgments. A lot of people, family, friends, colleagues, have been helpful, supporting and inspiring in the whole intellectual process that ultimately led to this book. In fact, it is already difficult to define the boundaries of that "intellectual process": what counts, and what does not, in all the thoughts, conversations, coffees, journeys, questions that occur during a period in which a book is being prepared? Some people will be left out from my list because of those "boundaries": I remember conversations that were not academic at all, and may even have been quite frivolous, but then after writing a given sentence in the book, I look back and find that the inspiration for that particular reflection came exactly from that conversation. So, a symbolic but big thanks go also to all these people I may have unfairly left out.

At any rate, even my attempt to thank the many remaining ones is far too ambitious, and one minute after these lines are published (mind you, one minute *after*, not before: Murphy's law is merciless, as we know) I will slap my forehead and shout "My goodness! How could I forget this person?". So, heartfelt apologies and another symbolic thank you to these people.

The others I can remember (both people and groups/institutions) are listed here in alphabetical order, without additional explanation, or measuring of their contribution. You know who you are, what you did, and you all know that whatever you did was very precious for me (the other sad part, a testimony to the days and society we are living in, is that my most reliable database to try and remember all these names was my contact list on Facebook. How decadent!)

Anyway, my deepest gratitude goes to:

Eleonora Adorni, Silvia Amodio, Oana Andreica, Kristian Bankov, Marcello Barbieri, Petras Baršauskas, Majid Behboudi, Aušra Berkmanienė, Miglė Bisigirskytė, Luis Emilio Bruni, Thomas Bryer, Eglė Butkevičienė, Catia Canciani, Marco Celentano, Nerijus Čepulis, Paulo Chagas, Davide Colonnari, Audronė Daubarienė, John Deely, Silvie Demandt, Teodoro Forcht-Dagi, Lucy Fleet, Mindaugas Gapševičius, Henrikas Gaidamavičius, Audronė Gedžiūtė, Eglė Gelažiūtė-Pranevičienė, Stefania Guerra, Kevin Holm-Hudson, the International

Semiotics Institute, Gisela Kaplan, Kaunas University of Technology, Vytautas Kederys, Seema Khanwalkar, Massimo Leone, the Lithuanian Academy of Music and Theatre, the Lithuanian Railways (the Vilnius-Kaunas train still being my favorite working place), Taylor Mali, Brian Marmion, Dainora Maumevičienė, Roberto Marchesini, Lina Navickaitė-Martinelli, Ricardo de Castro Monteiro, Juha Ojala, Alin Olteanu, Stephen O'Reilly, Roger Parent, Saulė Petronienė, Aida Pitkienė, Rima Povilionienė, Asta Pundzienė, Andrzej Raczkowski, Lina Rukevičiūtė, Farouk Seif, Jurgita Šiugždinienė, Springer, Rūta Stanevičiūtė, Simona Stano, Dalia Staponkutė, Gino Stefani, Aldona Steponavičiūtė, Peter Stockinger, my students in Kaunas, Helsinki and Rovaniemi, Natalya Sukhova, Eero Tarasti, Eila Tarasti, John Tredinnick Rowe, the University of Helsinki, the University of Lapland, the University of Torino, Jolanta Vaičiūnienė, Ulrika Varankaitė, Voices 4 Animals, Rokas Zubovas.

The only exception to this order shall be of course my own family, which I thank especially warmly and to whom this book is dedicated.

Kaunas, Lithuania Dario Martinelli

Contents

Chapter 1
Background: The Crisis of the Humanities

We are in the midst of a crisis of massive proportions and grave global significance (...) a crisis that goes largely unnoticed, like a cancer; a crisis that is likely to be, in the long run, far more damaging to the future of democratic self-government: a world-wide crisis in education.

(Nussbaum 2010: 1–2)

Abstract Humanities—it has been underlined by many sources—are currently facing a multilayered type of crisis that involve, at the same time (and in organic fashion), their impact on, and role within, society; their popularity among students and scholars; and ultimately their identity as producers and promoters of knowledge. The modern western world, with its primarily economic changes (often in a ruthlessly capitalistic direction) and the establishment of new priorities, has been identified as the strongest cause of such a crisis, not only *creating* the conditions for, but in fact encouraging it ("If the humanities are withering on the branch, it is largely because they are being driven by capitalist forces while being simultaneously starved of resources", as Terry Eagleton has it). While this is certainly true, to simply blame external factors is rarely a wise strategy, and a thorough, self-critical assessment of the situation is normally of great help for a smoother resolution.

1.1 The "Facts and Figures" Approach

The main goal of this whole chapter is provide an overview of the various ways the crisis of the humanities has been analyzed and assessed. As already specified in the abstract, the spirit of this overview is more self-critical than engaged in the type of sport that, among humanists, is definitely more popular than football: blaming the system. Provided the (many) faults that institutions and governments of any level (from entire countries to single academic institutions) have to be held responsible for, we have to admit that our primary fault as humanists was that of stubbornly thinking that, yes, the world could radically change *around*, but never really affect us: we felt that our identity was sacred, and yet our identity was exactly the thing we were betraying the most. I shall later elaborate on this (particularly in the second

D. Martinelli, *Arts and Humanities in Progress*, Numanities - Arts and Humanities in Progress 1, DOI 10.1007/978-3-319-45553-2_1

chapter of this book), but for the time being a brief overview of the various analyses of the crisis produced in the last few years is called for, in a way that mixes both self-critical statements with "blaming the system" shots.

In order to provide a 360° display of the various positions, it may be useful to bring in contributions from different modes of expression and genres of discourse: not only should scientific essays be reviewed, but also statistics, academic reports, public speeches, round-tables in conferences and newspaper articles. Each of them expresses different points of view, has different purposes, addresses different readers and considers different aspects and portions of the problem. All considered, I dare suggest that the analysis of the problem can be categorized in six main groups of approach, for which I have chosen a few specific references as a practical illustration.

The first group is based on numbers. The evidence of a decrease (occasionally loss) in interest towards humanities is documented in various ways. In Europe, the Instituto Nacional de Estadística has calculated a significant drop in degrees in humanities between 1998 and 2010. Specific institutes have also provided their own data: Harvard University has famously reported a straight 50 % cut in enrollments in humanities, from the 14 % of 1966 to the 7 % of 2010 ("Addressing a Decline in Humanities Enrollment", 2013); while 50 % is also the rough decrease that occurred at Yale, where we also learn that there is no significant difference between male and female students.

However, not all sources agree on this. Prof. Benjamin Schmidt from Northeastern University (Boston, MA), maintains that "the entirety of the long term decline [of the humanities] from 1950 to the present has to do with the changing majors of women" (Schmidt 2013). Before the 1970s, indeed, a remarkable percentage of women who went to college majored either in education (40 %) or in humanities (50 %). The second-wave of feminism, instead, encouraged women to pursue other paths, particularly business, math and science. Inevitably, the overall percentage of students in humanities suffered from it.

In addition to this, several universities in the whole world have joined social sciences and humanities together, under a single faculty (or department, or school), hoping to reciprocally strengthen programs and scholarships. Quite interestingly, nearly all of the marketing strategies of these institutions convey a similar message: humanities + social sciences = skills and values for a better world. Some examples: the tagline from the MIT School of Humanities, Arts and Social Sciences is "Great ideas change the world"; at Penn State Behrend College is "Developing the skills for success in a diverse world"; Jacobs University (Germany) suggests that the programs in their School of Humanities and Social Sciences have been "designed to combine elements of creativity, innovation, job perspectives and research orientation for an internationally competitive academic education"; Lakehead University inaugurates the page of the Faculty of Humanities and Social Sciences by quoting Socrates: "the unexamined life is not worth living"; and so on.

I can also mention my first-hand experience of this phenomenon, as my main academic affiliation, at Kaunas University of Technology, Lithuania, turned from a Faculty of Humanities into a Faculty of Social Sciences, Arts and Humanities (merging of course with the Faculty of Social Sciences) in 2013–14. As member of a faculty re-organization committee, I also witnessed in person (and creatively contributed to) the strategy of redefinition of identity. The decision was to use the tagline of "B.R.I.D.G.E." to market the faculty. That is: (1) an acronym for Brilliance, Research, Innovation, Democracy, Globality and Ethics; thus (2) a metaphor of how the faculty will now operate as an interdisciplinary connection across various disciplines; and, for the record, (3) a reference to one of Kaunas's main sightseeing attractions over the river Nemunas (a nice bridge, indeed).

More questions have been emphasized. In a remarkably exhaustive study called *The Humanities matter*, conducted by 4Humanities (a collective of American, Canadian, British and Australian scholars in Digital Humanities), it was shown that important positions of political and financial leadership in society are actually occupied, for a significant majority, by people with a humanistic background. As of 2012, 60 % of American CEO's, 65 % of UK parliament members and over two thirds of private entrepreneurs in wholesaling and retailing have degrees in humanities. In Canada, during 2008 (the year of the financial recession), 91 % of graduates in humanities were still able to find a job. 54 % of the economic importance of UK universities comes from humanities and social sciences (additional data are available at Terras et al. 2013).

Still, not more than 0.45 % of American federal research money and not more than 1.06 % of the EU research budget went to research in humanities, creating an enormous gap with other fields of inquiry. Of course, on the one hand, humanities are normally an inexpensive and relatively low-tech enterprise. While the nuclear engineer or the biochemist may need equipment and premises that are worth millions, the humanist's higher expenses are normally a few conference trips, some books and a new laptop. At the same time, that does not account for a difference of such proportions. Particularly when we think that, in fact, there are fields in the arts and humanities in which very costly equipment is needed. Building a decent music recording studio, complete with mixing desk, recording machines, Pro Tools, instruments, amplifiers, sound boots, microphones, and all the rest, is in fact an enterprise no cheaper than a biochemistry lab.

There must be something else.

1.2 The "Value of the Humanities" Approach

Most intellectuals predictably reacted with indignation to such data. Two of the best known examples are Martha Nussbaum's book *Not for Profit—Why Democracy Needs the Humanities* (2010), and a report from the American Academy of Arts and Sciences, called *The Heart of the Matter* (Broadhead et al. 2013). The focuses are slightly different (and, as I said, we on purpose want to examine different sources), but there is one important, common message: humanities should be supported because they create (1) "knowledge, skills, and understanding [needed] to thrive in a twenty-first-century democracy" (Broadhead et al. 2013: 10), (2) innovativeness, competitiveness and strength (ibid.: 11), and (3) "leadership in an interconnected world" (ibid.: 12). As Nussbaum put it:

> If the real clash of civilizations is, as I believe, a clash within the individual soul, as greed and narcissism contend against respect and love, all modern societies are rapidly losing the battle, as they feed the forces that lead to violence and dehumanization and fail to feed the forces that lead to cultures of equality and respect. If we do not insist on the crucial importance of the humanities and the arts, they will drop away, because they do not make money. They only do what is much more precious than that, make a world that is worth living in, people who are able to see other human beings as full people, with thoughts and feelings of their own that deserve respect and empathy, and nations that are able to overcome fear and suspicion in favor of sympathetic and reasoned debate. (Nussbaum 2010: 143)

The humanities can of course be defended and promoted in dozens of different ways, but I would not be surprised if most of us agreed that the most powerful rhetoric device employed in such promotions is the question of social/ethical values. And I mean rhetoric in the noblest sense of the term, not as a form of hypocrisy. I myself will use expressions such as "natural sciences study what makes life possible, humanities study what makes life worthwhile", later on in this book. Another one I heard is "natural sciences lead to better things, humanities lead to better people". And so forth: most of us think (and not incorrectly) that the humanities play a crucial role in the creation of a better, more democratic, more sensible society. The "value of the humanities" approach is therefore, very likely, the most common argument employed by humanists who discuss the crisis.

1.3 The "Humanities Need to Recover Strength" Approach

A loud minority, however, disagrees with this approach. Dr. Diana E. Sheets, from the University of Illinois, in a Huffington Post article (Sheets 2013), interestingly sees the crisis as a connection between the "[saturation] in the politics of social justice" and the poverty of "cultural references" in the humanities (ibid.). In other words, while humanists keep themselves busy with "identity politics, social struggle and gender warfare" (ibid.), they miss completely the great classics of the arts and literature, which are not just "great", but most of all challenging and demanding. The type of cultural knowledge they develop comes from *other* sources (the media most of all), and of course that can only consist of *popular* culture:

> [Students lack] the conceptual techniques to back up their opinions with authoritative evidence. Why? They [have] not been informed that some scholarly sources are better than others, that critical thinking means becoming knowledgeable and open minded about ideas that might be antithetical to one's beliefs. (ibid.)

While this happens, Sheets continues, scientists and engineers are doing and studying things that are exactly as difficult as they used to be, gaining a territorial advantage over humanists.

The solution?

> Our students must understand the foundations of their Western cultural heritage; they must be serious readers; they must engage in substantive research that necessitates the difficult process of critical thinking even if it challenges their worldview. This necessarily means returning to the foundational building blocks of Western civilization based upon rigorous standards of academic excellence. The humanistic disciplines must jettison the pseudo-Marxist catechism of social justice and postmodern nihilism exemplified by John Rawls in order that our students resume the pursuit of truth, beauty, and goodness, the firmament of the Western canon. (ibid.)

Altogether, this view is indirectly connected to the one expressed by the likes of Prof. Peter André-Alt (2012), from the Free University of Berlin, who also perceives humanities as the very tool for intellectual strength, profoundness and independence. Another example is Dr. Diana Sorensen, dean of arts and humanities at Harvard, who also maintains that the humanities help "to develop habits of mind, to develop a sense of how to reason rigorously, how to express ideas in a compelling way, and how to write well. And then to have at your disposal the categories through which to approach complex problems" ("Addressing a Decline in Humanities Enrollment", 2013).

Approaches of this sort are arguably the only ones that clearly go in a self-critical direction. So far (and in the next cases, too), we have only witnessed accusations towards "the system".

1.4 The "Humanities Are Not Really Science" Approach

The above-mentioned approach, too, has its own detractors. For instance, in his commentary to the crisis of humanities, Lee Siegel from *The Wall Street Journal* writes:

> We are told that the lack of a formal education, mostly in literature, leads to numerous pernicious personal conditions, such as the inability to think critically, to write clearly, to empathize with other people, to be curious about other people and places, to engage with great literature after graduation, to recognize truth, beauty and goodness.
>
> These solemn anxieties are grand, lofty, civic-minded, admirably virtuous and virtuously admirable. They are also a sentimental fantasy.
>
> The college teaching of literature is a relatively recent phenomenon. Literature did not even become part of the university curriculum until the end of the 19th century. Before that, what came to be called the humanities consisted of learning Greek and Latin, while the Bible was studied in church as the necessary other half of a full education. No one ever thought of teaching novels, stories, poems or plays in a formal course of study. They were part of the leisure of everyday life. (Siegel 2013)

There is often a tendency to identify humanities as a whole with literature alone, and this has to do with specific cultural perceptions (e.g., very often a separation is made between humanities and the arts, including *art research* in the latter), but, for the purposes of this discussion, let us just consider "literature" as a synecdoche for the whole field. Siegel also points out that it is "a blessing" for literature (=humanities) that "students choose to major in fields that are most relevant to the life around them" (ibid.), because now youngsters will be free to discover the classics for personal pleasure and education ("This socially and economically worthless experience is called transcendence, and you cannot assign a paper, or a grade, or an academic rank, on that. Literature is too sacred to be taught. It needs only to be read", ibid.). So, even if this was probably meant as a compliment, Siegel's suggestion is to keep humanities outside the constrictions of academic contexts and let them be "life lessons" instead.

1.5 The "BuBu" (Business and Bureaucracy) Approach

In the last decade or so, many researchers and teachers have also become concerned (and very frustrated) with the massive increase in managerial staff and management tasks within the academic world, which has led to a phenomenon called hyper-bureaucratization.

> Instead of government by academics there is rule by hierarchy, a good deal of Byzantine bureaucracy, junior professors who are little but dogsbodies, and vice chancellors who behave as though they are running General Motors. Senior professors are now senior managers, and the air is thick with talk of auditing and accountancy. Books—those troglodytic, drearily pre-technological phenomena—are increasingly frowned upon. [...] Philistine administrators plaster the campus with mindless logos and issue their edicts in barbarous, semiliterate prose. (Eagleton 2015)

Academic activity is nowadays monitored and measured by a multitude of rankings, applications, codes, reviews, evaluations that were previously non-existent and by now tend to occupy a good third, if not more, of the total working time of the average academic. In point 2 of the manifesto that I will present later in this book, I shall analyze in more detail this particular case, but for the time being it is important to remark that hyper-bureaucratization has certainly slowed down *every single* academic activity, not just those concerned with humanities, but perhaps humanists are in a particularly difficult condition here, as most of these managerial tasks are in fact conceived for types of measurements that are very difficult to apply to humanities (when not completely inapplicable).

Most of all, and predictably, the main problem is that all this "managerialization" of universities is aimed at transforming them into business enterprises where "academic merit is equated with how much money you can raise", where "professors are transformed into managers", and "students are converted into consumers" (ibid.). No wonder that the least profitable academic businesses, humanities, is the one that suffers the most from this situation, experiencing the most severe budget cuts (as we already emphasized in point 1 of this list), and the most pressure for redefinition of its research aims and education programs:

> Universities fall over one another in an undignified scramble to secure their fees. Once such customers are safely within the gates, there is pressure on their professors not to fail them, and thus risk losing their fees. The general idea is that if the student fails, it is the professor's fault, rather like a hospital in which every death is laid at the door of the medical staff. One result of this hot pursuit of the student purse is the growth of courses tailored to whatever is currently in fashion among 20-year-olds. In my own discipline of English, that means vampires rather than Victorians, sexuality rather than Shelley, fanzines rather than Foucault, the contemporary world rather than the medieval one. It is thus that deep-seated political and economic forces come to shape syllabuses. Any English department that focused its energies on Anglo-Saxon literature or the 18th century would be cutting its own throat. (ibid.)

Ladies and gentlemen: the knowledge economy!

1.6 The "Humanities Versus Technology" Approach

To conclude with a rather pessimistic view, more and more humanists see technology as the "great enemy" of the humanities. I shall devote specific portions of my own analysis to this particular point, so for the time being I shall just offer two quotes. One is from Leon Wieseltier, from *The New York Times* (2015), and it is one of the classic statements that many humanists like to repeat after they read *1984* or watched *2001: A Space Odyssey* for the first time:

> Beyond its impact upon culture, the new technology penetrates even deeper levels of identity and experience, to cognition and to consciousness. Such transformations embolden certain high priests in the church of tech to espouse the doctrine of "transhumanism" and to suggest, without any recollection of the bankruptcy of utopia, without any consideration of the cost to human dignity, that our computational ability will carry us magnificently beyond our humanity and allow us to transcend these limitations of our biological bodies and brains. There will be no distinction, post-singularity, between human and machine.
>
> [...] There is no more urgent tasks for [...] intellectuals and writers than to think critically about the salience, even the tyranny, of technology in individual and collective life. All revolutions exaggerate, and the digital revolution is no different. (Wieseltier 2015)

And the second quote is from Prof. John Paul Russo (1998), University of Miami, who, with a little help from Jacques Ellul's philosophy, closes the circle of this dystopian prophecy by suggesting a spiritual "personal self-transformation" as the solution:

> While this may seem far-fetched, it is worthwhile to recall that monastic refuge happened once in Western culture and the humanities survived. The light of learning at Lindisfarne and Cîteaux, at St. Gall and Monte Cassino, could be rekindled by an apprenticeship to the word in the midst of our necessary participation in technological society. As Burckhardt said, the culture of the West may once again be saved by ascetics. (Russo 1998: 41)

Thus, technophobia, too, is part of the problem.

1.7 Conclusions

How to summarize, then, the whole situation, and how to see the opportunities for an improvement? The various approaches so far listed have probably put an emphasis on three categories of analysis (see also Fig. 1.1):

1. **The general strengths of the humanities**: humanities promote critical thinking and analytical reasoning; they provide knowledge and understanding of democracy and social justice; they develop leadership, cultural and ethical values.
2. **The current problems of the humanities**: humanities lack economic relevance; they are perceived as "impractical" and unemployable; they do not match with technological development.
3. **The manifestations of the crisis**: little or no funding from institutions; a decrease in students; a decrease in interest; a loss of centrality to society.

This book will develop a project for "New Humanities", and in order to do so all these aspects should be considered. However—as mentioned at the beginning—self-critical assessment must not be forgotten, and in fact must be put forward, for the same reason why *every single* assessment of any problem should depart from self-analysis and self-criticism. Indeed, what has so far been exposed is the reasons why humanists should blame "the system", not why they should blame themselves:

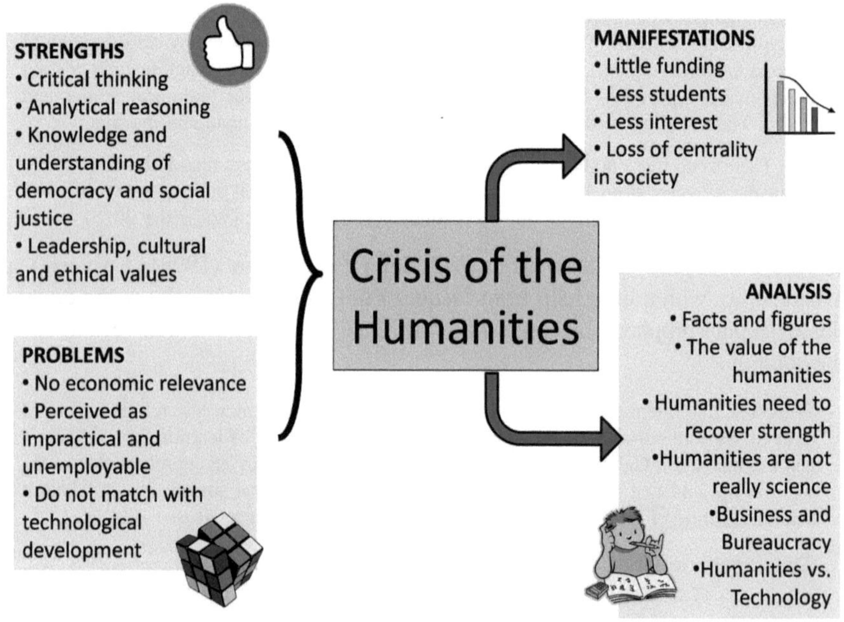

Fig. 1.1 The crisis of the humanities in bullet points

if we want to reform (or rediscover, or rescue, or else, depending on the viewpoints) the humanities, it is crucial to combine both aspects. Altogether, my arguments will be formulated in a manifesto of seven main "aims" that the humanities should rethink and pursue.

To employ a catchy expression, the umbrella-term "Numanities" (i.e., New Humanities) was coined, and under that label we have created a research concept that coordinates all the activities of my main affiliation, the International Semiotics Institute, at Kaunas University of Technology (www.isisemiotics.eu). It is important to specify this for two reasons: first because by the time this book comes out, there will have already been a few years of activity of "Numanities" as such, so this text should not be regarded as the "inauguration" of the concept, but rather its institutionalization on printed paper. Second, because one part of this book (namely its Appendix) will be devoted to an institutional history of the concept, and that of course will involve the International Semiotics Institute in the frontline.

The goal of Numanities, in synthesis (I shall of course elaborate on this aspect later) is to unify the various fields, approaches and also potentials of the humanities in the context, dynamics and problems of current societies, and in the attempt to overcome the above-described crisis. In fact, while "crisis" is the term that many scholars refer to, when they analyze the current condition of the humanities, one should also keep in mind that the word should be taken not only in its negative connotation, but in a wider, etymological sense (that is, the Greek *krisis*, which rather means "selection, judgment, turning point"). In other words, a crisis is a precondition for evolution, an opportunity.

Also, to discuss Numanities in terms of an "umbrella-concept" means also that there are no *specific* scientific contents in it, and that is why it is preferable to set the present document in the form of a manifesto, rather than a scientific paradigm or the likes. That particularly means that the many existing new fields and research trends that are addressing the same problems (post-humanism, transhumanism, digital humanities, etc.) are not *competitors* of Numanities, but rather possible ways to them (I shall discuss this aspect in the first point of the manifesto itself).

What we are launching here is not a new "theory", or "philosophical paradigm": to pursue "Numanities" does not mean to adhere to a particular research trend, but it is rather an array of "scholarly attitudes" of those researchers who are working on relocating and redefining the humanities. The goals set in the manifesto are hopefully a program, more than a specific theory.

Chapter 2
A Manifesto of Numanities

Abstract The present chapter, the actual core of this book, is an attempt to assemble a mission for Numanities. That is, what are the most important and urgent actions that a platform of this sort (with its intents to rediscuss the role, the challenges and the potentials of the humanities in modern society) need to take in order to reacquire that social and cultural relevance that seem to be crucial, in order to defend and promote critical thinking and analytical reasoning; provide knowledge and understanding of democracy and social justice; develop leadership, cultural and ethical values? At the current, initial, stage of this platform, the mission will be articulated in a manifesto of seven points.

2.1 Rethink the Position of the Humanities in Modern Society

> The humanities remind us where we have been and help us envision where we are going. Emphasizing critical perspective and imaginative response, the humanities – including the study of languages, literature, history, film, civics, philosophy, religion, and the arts – foster creativity, appreciation of our commonalities and our differences, and knowledge of all kinds. The social sciences reveal patterns in our lives, over time and in the present moment. Employing the observational and experimental methods of the natural sciences, the social sciences – including anthropology, economics, political science and government, sociology, and psychology – examine and predict behavioral and organizational processes. Together, they help us understand what it means to be human and connect us with our global community. (Broadhead et al. 2013: 9)

This lovely description of what the humanities are about should be a constant reminder for us throughout this whole monograph. We need to keep in mind what our identity is, but also how it can develop. And when we think about our identity, we may want to think of it in relation to the representation we are giving of it today. One of the suspicions that I maintain here in form of critical analysis is that in our struggle to preserve the role of the humanities in society made us lose the focus on what we were *really* defending. I shall elaborate on this further, but at this stage it is important to underline that, possibly, instead of interpreting our identity in that

D. Martinelli, *Arts and Humanities in Progress*, Numanities - Arts and Humanities in Progress 1, DOI 10.1007/978-3-319-45553-2_2

multifaceted, eclectic and dynamic manner that all identities imply by definition (particularly the identity of a whole area of inquiry that is formed by several disciplines and interconnections among them), we decided to crystallize it into a specific portion of space-time (which I would locate in the territory of European culture in a period between Romanticism and Post-modernism), and thus we ended up struggling for familiarity rather than for identity. Also, in doing so, we contributed to an unfair image of the humanities as a way to knowledge that is more content (in fact, self-indulgent—as Post-modernism often shows) with its limits than its potentials. The "keepers of the republic" lost the republic, because they wanted to keep it in a relatively small garden surrounded by fences.

For this reason, and regardless of the importance that they may acquire in present or future scenarios, it is crucial that the humanities (regain the right to) have a recognizable and authoritative role in research and everyday practices. The keyword, here, should be "dignity". If we compare the centrality of humanists in the past centuries with the marginality of nowadays, it is clear that the first crisis experienced by humanities, in order of moral priorities, is a crisis of "dignity". As I often tell my students: there used to be a time when if somebody asked you "what do you do in life?" and you answered "I'm a philosopher", or "I'm a poet", they would have looked at you in admiration as a prominent figure of the community, indeed a "keeper of the republic". Imagine the same answers now, the most you can expect from people is either "OK, sure, but what do you do *really*?", or, at worst, a compassionate pat on the shoulder which basically means "Oh, poor thing, so you are unemployed".

To recover dignity, the humanities must rethink their position and role at all levels, from the choice of research topics to the selection of the right platform to showcase them; from the approach to writing a project application, to the whole way they *read* and *interpret* the world.

In thinking of a "New Humanities" project, all these aspects should be considered, but—as mentioned at the beginning—self-critical assessment must not be forgotten. Indeed, what has so far been exposed has been the reasons why humanists should blame "the system", not why they should blame themselves. In this manifesto, and starting from this first point, I intend to combine both aspects, critical and self-critical ones. Altogether, the argument is formulated in seven main "aims" that the humanities should rethink and pursue.

The goal of Numanities is to unify the various fields, approaches and also potentials of the humanities in the context, dynamics and problems of current societies, and in the attempt to overcome the above-described crisis. In fact, while "crisis" is the term that many scholars refer to, when they analyze the current condition of humanities, one should also keep in mind that the word should be taken not only in its negative connotation, but in a wider, etymological sense (that is, Greek *krisis*, which rather means "selection, judgment, turning point"). In other words, a crisis is a precondition for evolution.

Also, to discuss Numanities in terms of an "umbrella-concept" means also that there is no *specific* scientific content in it, and that is why it is preferable to set the present document in form of manifesto, rather than scientific paradigm or the likes.

That particularly means that the many existing new fields and research trends that are addressing the same problems (post-humanism, transhumanism, digital humanities, etc.) are not *competitors* of Numanities, but rather possible ways to them. What we are launching here is not a new "theory", or "philosophical paradigm": to pursue "Numanities" does not mean to adhere to a particular research trend, but it is rather the array of "scholarly attitudes" of those researchers who are working on relocating and redefining humanities. The goals set in the present manifesto are hopefully a summary of the *attitudes*, more than specific theoretical proposals.

There are brand new ideas about how the humanities can be modernized, refreshed, "rescued" from the crisis, revitalized, and they are all significant to our project (and book series related to it). I do not intend to go through all of them (also because I have no idea of how many I would ignore in such a hopeless attempt), but I can mention a few of those I have met during the research conducted for this book, which, like Numanities, (a) have approached the issue of the humanities' reformation with a programmatic agenda (including other manifestos), (b) are also research concepts within an academic institution (like Numanities is for the International Semiotics Institute, at Kaunas University of Technology), and ultimately (c) have the fundamental goal (and spirit) of "rethinking the position of the humanities in the current society". I shall proceed to mention these cases in no particular order.

Of great interest is certainly the project of the transformative humanities, which, as we shall often emphasize here, makes no mystery on the fact that

> [...] the humanities of today are enveloped in the paradigms of the past. Even the most recent attempts to deal with the state and prospects of the humanities in the twenty-first century largely explore the socio-political and educational aspects of humanistic teaching and research in the academy, the place of the humanities in the curriculum, and their interaction with the natural and social sciences in the infrastructure of a university. Many of such attempts are devoted to historical ideas in the humanities and do not have much to say about their path to the future. (Epstein 2012: 10)

On the contrary, Epstein sees a great opportunity for the development and redefinition of the humanities in the current historical epoch, since "for the first time in history, humans now have the ability to create something (or even somebody) similar to humans themselves. Not just tools or symbols, but artificial intelligence, artificial organisms, new forms of life, and holistic human-like creatures" (Epstein 2012: 11), urging humanists not to miss the opportunity to perform "their own attempt to enhance the role of humans in the transformation of the universe and their own nature" (ibid.)

Epstein's agenda, pursued in his Center for Humanities Innovation at Durham University, is mostly identifiable with the wider paradigm of transhumanism,[1] and

[1]We shall leave to another discussion the controversies that have occasionally arisen around this notion, like the famous warning written by Francis Fukuyama in the special issue of *Foreign Policy* on the "The World's Most Dangerous Ideas" (published in 2004). On that occasion, transhumanism enjoyed the company of none other than "The War on Evil", "Business as usual at

revolves around eight main objectives (Epstein 2012: 293–294): (1) "To focus specifically on the potential of the humanities to transform the objects of their study"; (2) "To address problems of technical enhancement and biological transformation of humans, simulated reality and artificial intelligence from the humanistic perspective"; (3) "To [have] the capacity to respond swiftly to significant new trends or intellectual problems that […] call for establishing new fields of research"; (4) "To develop new disciplines in the humanities that respond to the socio-cultural challenges of the 21st century"; (5) "To expand the genres of intellectual discourse, with particular emphasis on those creative and concise genres [like] manifestos, theses, aphorisms, fragments, programs, ideas and notes"; (6) "To examine how new informational technologies radically change"; (7) "To develop specific methods and criteria for the evaluation of the transformative power of thinking and intellectual creativity, imagination, inspiration, invention and originality in the humanities"; and (8) "[To develop] InteLnet—an electronic portal for intellectual innovations" (it can be found at http://www.emory.edu/INTELNET/intelnet2.html).

Another project that is extremely welcome in the forum of Numanities was generated in the environment of the Hollandsche Maatschappij der Wetenschappen, the "Dutch Society of Arts and Science", and is best encapsulated in Tindemans et al. (2003). The main emphasis here is interdisciplinarity, considered an unavoidable destiny of modern research (and a priority for Numanities, too, as we shall soon see): "The future of every bit of science depends on the ability to use all instruments from the toolbox, whether they have their origin in and were even largely identified with the natural sciences on the one hand and the social sciences and the humanities on the other" (Tindemans et al. 2003: 13). The following are the main conclusions:

- Disciplines will remain with us, but no longer as the primary basis for research arrangements.
- Naming domains of heterogeneous scholastic enterprise may present creative challenges, but reference to the constituents and the specific tools will help explain what such innovations are about.
- Successful new departures—often embedded in and triggered by primarily 'engineering' type organizations—will require the support of more than one sector of society. There should not just be an industrial background or, conversely, exclusive government funding.
- Openness is what is most needed. Tribal behaviour, a characteristic of many existing research establishments, is to be avoided.
- Funding should be given wisely. As one discussant said, "what research needs is not just a budget, not just a focus, but both simultaneously".
- It is also obvious that science—in the wider sense of including studies in the humanities and social sciences – must be truly embedded in society. (Tindemans et al. 2003: 15–16)

(Footnote 1 continued)

the U.N.", "Free money", "Undermining free will", "Spreading democracy", "Religious intolerance" and "Hating America". Not bad.

Next, I shall like to mention the project "New Directions in Humanities", which revolves around the website thehumanities.com and features three main outlets: "The International Conference on New Directions in the Humanities" (an annual event that, at the time I am writing these lines, has reached the 13th edition, the 14th being planned in 2016 at the University of Chicago); a rather rich variety of publications, including seven peer-reviewed journals and a book series published by Common Ground Publishing Books, consisting so far of about 40 titles; and a web community called "New Directions in the Humanities Knowledge Community" (where members have numerous chances to be involved, sharing video presentations, participating in boards and editing work in the various publications; the community is also on Facebook, Twitter, and Scholar).

The project was born in 2003 with the purpose of "developing innovative practices and setting a renewed agenda for [the] future [of the humanities]" (http://thehumanities.com/about). As with Numanities, here, there is no specific attempt to establish a new paradigm in the humanities, but rather to collect a number of them under a single umbrella. Very interestingly, their scopes are grouped under three categories of relations: Humanities-Science-Technology; Humanities-Economy-Commerce; and Humanities Themselves (http://thehumanities.com/about/scope-concerns). In the first case:

> The new technologies and sciences of informatics […] are infused to a remarkable degree with the human of the humanities: the human-centered designs which aim at 'usability'; the visual aesthetics of screen designs; the language games of search and tag; the naming protocols and ontologies of the semantic web; the information architectures of new media representations; the accessibility and manipulability of information mashups that make our human intelligence irreducibly collective; and the literariness of the code that drives all these things. So too, new biomedical technologies and sciences uniquely inveigle the human—when considering, for instance, the ethics of bioscience and biotechnology, or the sustainability of the human presence in natural environments. (http://thehumanities.com/about/scope-concerns)

In the second case (Humanities-Economy-Commerce):

> In the modern world, 'economy' and 'production' have come to refer to action and reflection pertaining to the domains of paid work, the production of goods and services, and their distribution and market exchange. At their etymological source, however, we find a broader realm of action—the realm of material sustenance, of domesticity (the Greek 'oikos'/household and 'nemein'/manage), of work as the collaborative project of meeting human needs, and of thrift (economizing), not just as a way of watching bottom lines, but of conserving human effort and natural resources. Today more than ever, questions of the human arise in the domain of the econo-production, and these profoundly imbricate human interests, needs and purposes. Drawing on the insights of the humanities and a renewed sense of the human, we might for instance be able to address today's burning questions of economic globalization and the possible meanings and consequences of the 'knowledge economy.' (http://thehumanities.com/about/scope-concerns)

Finally, when it comes to the humanities themselves:

> [...] the humanities are considered by their critics to be at best esoteric, at worst ephemeral. They seem to have less practical 'value' than the domains of techno-science and econo-production. But what could be more practical, more directly relevant to our very existence than disciplines which interrogate culture, place, time, subjectivity, consciousness, meaning, representation and change? [...] Within this highly generalized scope, [we] have two particular interests:
>
> - Interdisciplinarity: The humanities is a domain of learning, reflection and action which require dialogue between and across discipline-defining epistemologies, perspectives and content areas.
> - Globalism and Diversity: The humanities are to be considered a space where recognizes the dynamics of differences in human history, thought and experience, and negotiates the contemporary paradoxes of globalization. This serves as a corrective to earlier modes of humanities thinking, where one-sided attempts were made to refine a singular essence for an agenda of humanism.
>
> The humanities come into their own in unsettling spaces like these. These kinds of places require difficult dialogues, and here the humanities shine. It is in discussions like these that we might be able to unburden ourselves of restrictively narrow knowledge systems of techno-science and econo-production. (http://thehumanities.com/about/scope-concerns)

In the light of these concerns, "New Directions in Humanities" declares to pursue the following areas of inquiry (http://thehumanities.com/about/themes): Interdisciplinary and multidisciplinary humanities; The relationship of humanities to other knowledge domains (technology, science, economics); Making knowledge: research in the humanities; Subjectivity and objectivity, truth and relativity; Philosophy, consciousness and the meanings of meaning; Geographical and archeological perspectives on human place and movement; The study of humans and humanity, past and present; The future of humanities.

Additional insights on the (inter)disciplinary contents of this community are detected in their publications and conferences. For instance, the titles of their seven journals are *Literary Humanities*; *Book*; *Civic, Political and Community Studies*; *Communication and Linguistic Studies*; *Critical Cultural Studies*; *Humanities Education*; *Humanities: Annual Review*.

To conclude, as with a previous footnote on transhumanism, I shall like to avoid here the controversial aspects of some of these enterprises (because I am mostly interested in emphasizing their existence, contents and similarities with Numanities). However, it must be said that in some online forums (see, for instance, *Scholarly Open Access*, which offers the very useful service to list the so-called predatory open access publishers) "New Directions in Humanities" has been reported as a possible predatory environment, particularly its conference which, according to these criticisms, is not very transparent. Having said that, these questionable aspects are very far from "certified" predatory publishers and massive spammers like David Publishing Company, World Academic Publishing and the likes (for a full list, I very much recommend the above-mentioned service at the page http://scholarlyoa.com/publishers/. I take the public opportunity here to thank them for the great hand they lend to the scholarly community).

As a fourth example, we cannot overlook the amazing job that the Massachusetts Institute of Technology is doing in creating a recognizable role for the humanities within the context of a technological institution (it is of course an issue I feel a particular interest in, as the International Semiotics Institute is currently in the same condition—a humanistic unit within a technological university). By employing the "value of the humanities" approach (their slogan is "Great ideas change the world"), MIT have assembled one of the most complete, eclectic and yet coherent units of humanities in the whole world. The MIT School of Humanities, Arts, and Social Sciences, one of five schools of the Institute, has 13 departments and grants S.B., S. M., and Ph.D. programs. Their mission is embodied in the word "Empowering":

> Generating practical solutions for the world's great challenges requires technical and scientific creativity—and an understanding of the world's human complexities: political, cultural, and economic. MIT's Humanities, Arts, and Social Sciences fields empower young scientists, engineers, thinkers, and citizens—with historical and cultural perspectives, and skills in critical thinking and communication—to help them serve the world well, with innovations and lives that are rich in meaning and wisdom. (http://shass.mit.edu/mission)

Within this framework, three areas of focus are identified: "Research and Innovation", with currently 21 research projects with topics as varied as "Repairing America's Social Safety Net", "How do societies confront histories of injustice?", "Production in the Innovation Economy", "Why are some nations wealthy while others are poor?", "How did slavery shape American universities?", and so forth; "Teaching Critical Skills", with 21 fields of study (alternating "classic" fields like linguistics, music and philosophy, with more specific/innovative ones like Comparative Media Studies, Science Writing and HyperStudio); and finally "Educating Leaders and Global Citizens", an area that mostly revolves around the famous MISTI Program (MIT International Science and Technology Initiatives—http://misti.mit.edu/), whose mission is the creation of practice-oriented, international learning experiences for MIT students in relation to their course of study, and in collaboration with foreign companies, universities and research institutions (22 partner-countries altogether).

Thematically, all the activities are grouped into the so-called "Five great challenges" (http://shass.mit.edu/mission/great-challenges): Innovation/Creativity; Social innovation; Education; Environment; Health/Health care.

It is of course pleonastic to praise such a big and rich institution like MIT, but one has to do it anyway, as their offer in humanities is something to look up to, while searching for an effective, innovative-yet-historically-conscious, top-quality integration of humanistic research into society and interaction with science and technology.

To conclude—*dulcis in fundo*, as the Latins would have it—we cannot ignore the big movement of posthumanism and its various branches (including transhumanism itself, which some people treat as a "sub-genre" of posthumanities—although that would not be my case). When we started the project of Numanities at Kaunas University of Technology, I was of course involved in several discussions (mostly friendly ones) on what Numanities were, and how they were related to humanities as such. A very recurrent question I would receive was "So… is this a sort of posthumanities?". I am not sure why I was specifically asked this question, but I have three hypotheses, all of them revealing something about the solid connection between the two concepts. First, I may have been asked this question because both the prefixes "New" ("Nu") and "Post" suggest an idea of advancement (at least in time, if not in contents). Second, because of all the various movements of reformation of the humanities, posthumanities is the one that is heard more often, in both scholarly and everyday environments (not to mention the artistic ones, where the word is particularly trendy, and has generated plenty of ideas and projects of connection between artistic and scientific practices). The third hypothesis (the most important, although—I fear—not many of those who asked me that question were in fact thinking about this): someone may have recognized a crucial and foundational common goal between Numanities and posthumanities. That is: they are both very committed to counterbalancing the dominance of subjectivism in the humanities with an emphasis on (a) non-human agents, other animals in particular; and (b) social practices (as opposed to individual ones). In her introduction to posthumanism (and to a whole book series on it, published by University of Minnesota Press), Cary Wolfe defines the field in the following way:

> My sense of posthumanism is [that] it comes both before and after humanism: before in the sense that it names the embodiment and embeddedness of the human being in not just its biological but also its technological world, the prosthetic coevolution of the human animal with the technicity of tools and external archival mechanisms (such as language and culture) […] and all of which comes before that historically specific thing called "the human" […]. But it comes after in the sense that posthumanism names a historical moment in which the decentering of the human by its imbrication in technical, medical, informatic, and economic networks is increasingly impossible to ignore, a historical development that points toward the necessity of new theoretical paradigms (but also thrusts them on us), a new mode of thought that comes after the cultural repressions and fantasies, the philosophical protocols and evasions, of humanism as a historically specific phenomenon. (Wolfe 2010: xv–xvi)

Note particularly this attention for the "social" (which we find elsewhere too), but mainly this decentralization of focus from the "human" to the "not-only-human", is something that we do not find in any of the other strategies of reformation for the humanities, neither the ones I mentioned above nor others I came across. Posthumanism is the main (if not the only) paradigm that takes this aspect into serious consideration. For Numanities, too, this is a crucial point, both from an epistemological and an ethical level. If an important goal behind the reformation of the humanities is the establishment of a fruitful interdisciplinary cooperation with natural sciences, then it is high time to open up to what science tells us about the continuities of humanity-animality, and how the former is not *alien* from the latter but simply part of it, just as *Homo sapiens*, like it or not, is an animal species. Wolfe is very clear on the role and the importance of the "non-human" in the posthumanities:

> What began in the early to mid-1990s as a smattering of work in various fields on human-animal relations and their representation in various endeavors — literary, artistic,

> scientific — has [...] galvanized into a vibrant emergent field of interdisciplinary inquiry called animal studies or sometimes human-animal studies. [...] I want to suggest that both rubrics are problematic in light of the broader context in which they must be confronted — the context of posthumanism. More specifically, [...] the questions that occupy (human-) animal studies can be addressed adequately only if we confront them on not just one level but two: not just the level of content, thematics, and the object of knowledge (the "animal" studied by animal studies) but also the level of theoretical and methodological approach (how animal studies studies "the animal"). Just because we direct our attention to the study of nonhuman animals, and even if we do so with the aim of exposing how they have been misunderstood and exploited, that does not mean that we are not continuing to be humanist — and therefore, by definition, anthropocentric. Indeed, one of the hallmarks of humanism [...] is its penchant for that kind of pluralism, in which the sphere of attention and consideration (intellectual or ethical) is broadened and extended to previously marginalized groups, but without in the least destabilizing or throwing into radical question the schema of the human who undertakes such pluralization. In that event, pluralism becomes incorporation, and the projects of humanism (intellectually) and liberalism (politically) are extended, and indeed extended in a rather classic sort of way. (Wolfe 2010: 99–100)

Not by coincidence, one of the case-studies for Numanities I will present in this book is closely related to this approach, so brilliantly summarized by Cary Wolfe. Not by coincidence, on the scientific board of this book series we have Dr. Gisela Kaplan, a world-wide known ethologist, and Dr. Roberto Marchesini (the leading Italian anthrozoologist, who is also a director of the Center of Posthuman Philosophy in Bologna), plus the researcher Eleonora Adorni (from the same center) on the editorial board. And not by coincidence, finally, a monograph by Marchesini is planned as one of the first publications of the series.

Now. All these inputs from other fields that are gathered here also represent the state of the art of a movement that is still taking its relatively-first steps (some fields are older, some are younger, but we are anyway talking about paradigms that have been developing from late 20th century at their earliest). Such a picture cannot of course be homogeneous, and it may even be self-contradictory, at times (or simply contradictory, as there is no "self" yet, in a field that is indeed looking for a recognizable identity). So, it is our specific intention, in the project of Numanities, to represent the heterogeneity of the paradigm, in this initial stage of its life: heterogeneity that also means conflict, or at least confrontation, in some cases. Reasoning in terms of this book series, the readers should not be surprised if they will witness a given valid argument in one publication and an equally-valid counter-argument in the next one.

Numanities is not a discipline: it is a forum, a platform, an agora. We have a manifesto because every forum has its rules and etiquette, but these are not meant to affect the development of the discussion. Rather, they are meant to trace the characteristics of the "spirit" that animates the discussion, and—occasionally—to warn against going too much "off-topic".

In the conclusions to this manifesto section, I shall propose a general list of "on-topic" humanistic areas of inquiry that are particularly relevant for this forum.

2.2 Reestablish the Dialogue Between Humanities and Institutions

A major problem in the current crisis of the humanities is that the latter and their institutional interlocutors (establishment, funding sources, management…) have ceased interacting in a fruitful manner. This communication breakdown can be particularly understood by considering two of the six types of analysis of the crisis that I have mentioned before: the "value of the humanities" approach and the "hyper-bureaucratization" approach. In the first case, it has become clear that institutions do not seem to appreciate the importance of the "uncountable" value created by the humanities. For instance (see also the "facts and figures" approach), we have seen that only a small, insignificant portion of public funds are allocated to humanistic research. In the second case, we have mentioned that the bureaucratic/managerial apparatus surrounding (in fact, swallowing) the academic world, is particularly designed to cut out any humanistic mentality or problem-solving skill.

Now. The spirit of this manifesto is first and foremost that of taking a thorough self-critical attitude before being allowed to proceed to criticize "the system". Therefore, we shall start by declaring, rather firmly, that although few commentators will admit it, it is up to the humanities to make a serious effort to learn the "rules of the game" of modern society. Complaining should be allowed only when all other (proactive) strategies have been attempted. The sensation, at least when it comes to the majority of humanists commenting on this status quo, is that there is a pinpointed analysis of the system's faults on one side, and a clear superficiality in pondering their own mistakes, inabilities and inactivity in the process. Most of all, nobody wants to admit that if we reached this particular situation, it is also (or perhaps mostly) because the academics have been, for decades, models of laxity, short-sightedness, recklessness and laziness. In more prosperous financial times, they just worried about themselves, their own position and their ego; failed to pave any path for future generations; used the general lack of institutional control for their own profit (instead of celebrating the value of that freedom), and did not care to treat their own research/teaching activities in terms of intellectual and economic investment. Humanists, unfortunately, have been in the front line in such an attitude, happy that the humanities did not need to be "justified" at an institutional level, as they enjoyed that "intrinsic value" that nobody had the courage to question. Except that now, guess what, someone had that courage, and indeed humanities lost that centrality that I mentioned in point 1 of the manifesto.

Totally taken by surprise by this frontal attack to their authority, humanists could not find a better solution than basking in self-pity for being now marginalized by the institutions, instead of facing the current difficulties by doing what they do best: reflect, analyze and look forward. Institutions are granting less and less funds to humanities, because they sense a lack of concreteness and profitability of the field, while humanists think that, if their research was concrete and profitable, it would simply not be humanist.

This, to begin with, is a historically-inaccurate assumption: tons of humanistic research and output were conceived with an empirical, foresighted and applied mind, profit included. Examples are many, and very famous. We could perhaps mention Filippo Brunelleschi, and remember that he was not hired to work on his famous dome (the "cupola") by some aristocrat or patron. He had to win a competition held by a "corporazione", the Arte della Lana (the wool merchants' guild), which, in 1418, held a contest to solve the old problem of the dome in Santa Maria del Fiore, Florence's then new cathedral. Brunelleschi had to compete with sixteen other colleagues, the most prestigious of which was another great artist of the period, Lorenzo Ghiberti. He won, and he got the project, delivering to history a most innovative architectural solution, the then-world's biggest dome (and still nowadays the biggest made with stone), and one of the most recognizable symbols of Florence.

Or, if we want more general examples, it should be enough to remember that philosophy (in different parts of the world, independently) was born with the specific aim of developing concepts of moral conduct, political decision-making, healthy lifestyle, social interaction, educational processes… the nowadays-typical idea of the absent-minded, detached-from-life, unaffected-by-daily-matters philosopher could not be further from the original state of things, and was, in fact, a relatively recent invention (we shall see that later, in point 3 of this manifesto).

In order to regain centrality in society, an imperative for humanists is to moderate the complaining and self-pitying part and reinforce the knuckle down part.

And, let us make no mistake, this does not only apply to "funding applications" and the likes: it is the whole *social image* of humanists that can use a bit of restyling. The concept of "intellectual" has somehow split into two separate categories that do not communicate with (and in fact despite) each other: an anti-establishment "bohemian outcast" on the one hand, and the business-oriented "manager" of knowledge on the other. The former has only the talent, the latter only the money, much to the detriment of cultural and intellectual progress. What got lost in the process is the "architect" of knowledge: the intellectual who combines the vision with the engineering, the abstract with the concrete. Some humanists seem to think that such a figure would be entirely new and inappropriate for humanities: on the contrary, this is the most ancient humanist there is.[2] This transformation has of course many causes, and this is not the place (nor do I have the competences) to deepen the issue. However, one thing that looks pretty clear to me—at least on a general level—is that, in one way or another, we all got trapped

[2]And, by the way, to all the liberals who are at this point shocked by these remarks on bohemianism (and will be even more later on): have you noticed that bohemian is the *new conservative*? Political groups, Facebook pages, blogs and associations self-styled as "Conservative bohemians" are popping up like mushrooms all around the world, redefining the ideological agenda of a social category that has been normally associated to the "left-wing" environment. Then again, when we consider that bohemianism has individual freedom and spirituality among its main pillars, that should hardly sound like a surprise.

into the myth of the "Romantic genius". The intellectuals and the artists who inhabit our cultural discourse (suffice to think about filmic representations of painters or composers) have to be tormented, instinctive and rigorously outside/against the establishment:

> On the one hand, the Romantic genius was a brooding, sensitive loner tormented by doubts and condemned to be misunderstood, and on the other, an inspired poet-prophet whose sublime enthusiasm for *le bel ideal* leads to empowered and impassioned eloquence. (Waller 1992: 224)

The "genius" cannot be a quiet, rational, pragmatic person. It is quite outstanding, and yet unsurprising, that great minds like Leonardo or Bach were never subjects of a proper biographical movie: endless documentaries, a few TV portrayals, and of course a lot of references in various productions, but no one so far has had the nerve to create anything remotely comparable to movies like *Amadeus* or *Bird*, where not only the two protagonists, Mozart and Charlie Parker, are given the full Romantic treatment of troubled and supernaturally-inspired personalities, but they are also assisted or rivaled by "quiet" talents (Salieri and Gillespie, of course), whose role, among other things, is that of showing the limits of a rational (that is, ordinary) approach to art. Someone like Bach is a contradiction, in this respect: a composer should either display the stereotype of the childlike, explosive eccentricity of Mozart, or the other stereotype of the stormy, impetuous passion of Beethoven. If you *regularly* sit every single week to write a one-hour cantata (like Bach did in his Leipzig period), ending up writing 200 of them (none of which standing below the threshold of the masterpiece), and in the meanwhile you are also a devoted family man, and a very diligent professional who seeks for the perfection of his job and not for fame and glory—then we have a problem. We do not have anymore that mix of *maudit*, drama and magic that the *cliché* requires: we have instead rationality, sobriety and orderliness—something much less media-sexy.

The humanities, perhaps, should make an effort to emancipate from this Romantic cliché, and understand that another way of staying in society is possible. A successful dialogue with institutions is not automatically a successful dialogue with society as a whole, but it is a good step in that direction.

Now, for the same reason we do not want humanists to be just "bohemian outcasts", we also do not want them to turn entirely into "managers of knowledge": what we want, let me repeat, is a middle that in this case could not be more golden. Therefore, some specific aspects of this advocated dialogue can and should be also achieved by a clever and strategically-organized (again: less complaining and more action) process of reformation, if not revolution. And that articulated group of practices that we have called "hyper-bureaucracy" is, in my humble opinion, *entirely* on the wrong side of this dialogue, and as such it must be profoundly changed, if not—as I would sincerely hope—disintegrated.

I shall begin my analysis with three quotations. First: "[…] As to the demands from the Faculty, that is what they typically do nowadays: disturb quiet people". This one, the softest, is a passage from an email written by my former PhD supervisor (and mentor and friend) Prof. Eero Tarasti to one of his students who was asking him about a new system of assessment of doctoral studies introduced by the University of Helsinki.

Second quotation:

> Imagine that managers are going to assess the quality of restaurant meals but they have no sense of taste. They have no idea – everything tastes the same to them. So what are they going to do? They will undertake evaluations such as how many minutes did it take for the soup to arrive at your table? How many words of explanation did the waiter use? And so on. Everything is evaluated quantitatively, so the obvious thing for a manager to do is to increase the amount of information gathered. As all these factors come into play, then you need more and more managers and managers need to cover themselves so they bring in more managers. So what happens is hyper-bureaucracy. My view is that this has its own momentum and there can be no limit to it and it is not something that can be rolled back. I would predict that the 33% expansion will be greater in 10 years' time because there is no mechanism that can put a stop to this. (Prof. Grahame Lock, philosopher at Oxford University, as quoted in Tahir 2010)

Third quotation:

> Universities are increasingly populated by the undead: a listless population of academics, managers, administrators and students, all shuffling to the beat of the corporatist drum [...] Academic zombie speech is peppered with affectless references to DEST points, citation indices, ERA rankings, ARC applications, esteem factors, FoR codes, AUQA reviews and the like. Aca-zombies participate in numerous Rber-zombified, government-sponsored quality assurance exercises, presided over by powerful external assessors. Many zombies have long lost the capacity to distinguish between a place of learning and a money-making PR machine, mummified in red tape. They appear incapable of responding meaningfully to the tyranny of performance indicators, shifting promotion criteria, escalating workload demands and endless audits, evaluations and reviews. [...]
>
> They even come to believe corporatist language promotes transparency and accountability. The viral effects of such delusions are such that many aca-zombies do not even realise they have already passed over into the valley of shadows. Work formerly conducted at university (remember teaching and research?) has been replaced by a sinister doppelganger: bureaucratically generated compliance.
>
> The virus is also present among the overworked reserve army of sessional minions, trapped in a stygian netherworld of precarious short-term teaching contracts. This legion of lost souls is the raw material for University Inc, a sinister operation similar to the Umbrella Corporation of the Resident Evil franchise. Like the undead slaves of voodoo lore, these tutors, entranced perhaps by a misapprehension regarding the status of full-time academic zombies, are mercilessly exploited but expected also to continue producing and publishing scholarly research. [...]
>
> Occasionally it is necessary, as in Zombieland and Shaun of the Dead, to pass as undead to survive. Paradoxically, it is the unthinking intellectual rigor mortis of the present bureaucratic plague that enables some to survive the worst aspects of zombification. (Gora and Whelan 2010)

There we go. In a crescendo of annoyance, frustration and fury, researchers and teachers (that is, in the metaphors, quiet people, restaurant chefs and finally human survivors in a land of zombies) display their concern over the massive increase of managerial staff within the academic world, and that phenomenon known as hyper-bureaucratization. Such staff overflows request all sorts of things from academics: monitoring, assessments, deconstruction and reconstruction of programs according to new rules, measuring and quantitative analyses. The cunning promise of a simplification of the procedures in the near future ("once you have gone through the learning curve, you'll see that everything will proceed faster", or—more diabolically—"this is just an evolution, not a revolution, from what you were already doing") is constantly deluded by repeated additions, so-called refinements, and most of all reshuffling and addition of more managers, who—in turn—are eager to introduce new procedures and new training sessions to teach them.

These people (Prof. Lock correctly remarked) have no competence in the academic work: they are not scholars-turned-to-managers, nor managers with *some* competence in scholarship. They are managers-full-stop. They may have worked in public administration, in a bank, in the best of cases in a governmental institution related to cultural policies, but that is about it. You quote Homer to them, and they happily reply that they also watch *The Simpsons*. The result is a particular case of the famous Putt's law (Putt 2006: 159): there are people (academics, in our case) who understand what they do not manage, and people (the bureaucrats) who manage what they do not understand.

Now, academics are already very allergic to any form of intrusion into their work (especially comments on its quality) when such intrusion comes from an experienced colleague. If it comes from people they do not even acknowledge as "equals", they get seriously irritated, and not without reason (why should a researcher be asked to perform administrative tasks by *administrative staff*? Is by any chance an administrator ever asked to write and publish—A-level, of course!—a scientific article?). Humanists are in the worst condition in this situation, because not only they do not acknowledge any authority in managers: they also do not understand a word of what they are told by them. While researchers and teachers in natural sciences will at least understand the numerical aspects of the tasks that are inflicted on them by managers (say, a financial plan, a budget statement…), and those in social sciences will get the methodology and the language, humanists will get frustrated with both figures and words, and that will anger them even more.

Not that one should be surprised, by the way, as humanists were not involved in this process in the first place. The whole managerial/bureaucratic system was conceived and developed within a territory (economics, social psychology, statistics, marketing research…) mostly inhabited by social and natural scientists, and where humanists were carefully left out of the picture. In that territory, for instance, the classic writings on bureaucracy and management (Marx, Mill, Weber… competences that humanists would have been eager and competent to contribute) are happily ignored, in favor of dynamic workshops, games of social facilitation, and of course plenty of markers and post-it's. Humanists are not welcome in the creation of this apparatus, not only because their competences are not perceived as necessary, but also (mostly?) because they are the ones most likely to understand the basic fraud behind it: while bureaucracies are rhetorically advertised as the only way to ensure transparency, reciprocal control, fairness, etc., they are in fact the antithesis to participatory democracy.

And why is it humanists who can understand this? Exactly because they, on the contrary, *did* read their Weber, and they feel that "polar night of icy darkness and hardness" right inside their bones (I am intentionally giving no reference here: humanists know what I am talking about, while managers, for a change, may finally be prompted into reading the classics of management theory—or at least Googling the very expression I mentioned!).

Anyway: what are the consequences of hyper-bureaucratization, and how do they affect the humanities in particular (provided that, as we said, it is the whole of academia that is affected by this process)? I will enter into details in the "manifesto" section of this book (particularly point 2), however a general list can be compiled here (the first three already emphasized by Weber himself[3]):

1. Too much (and most of all, too redundant) red tape and paper work: action, decision-making and problem-solving, instead of being facilitated, are slowed down or even hindered. Humanists, as mentioned, suffer in particular for this, as they also need additional time to adapt to a *forma mentis* they were never familiar with in the first place. In one sentence: they are exactly asked to renounce to what they are best at.

The tragic case of the procurement system is possibly the most painful example of red tape delirium. Born to avoid lobbying and conflict of interests in the various service and material expenses that an academic institution faces, this "best of three offers" principle has become a neurotic practice that not only slows down and complicates procedures that once were very quick, but also determines a standardization and worsening of the service itself. First of all, a company owner who wins a procurement for—let us say—printing a book, knows already that s/he will get the money *before* printing, so the healthy principle of competition that was previously established in the tender with two other companies is threatened by the serious possibility that the service eventually delivered will not be as good as promised (as it so often happens with services paid in advance): knowing very well that the bureaucratic apparatus of the university is so laborious that the chances anybody will raise an issue (including legal means) are very low, the company makes a very clear point on why it was the cheapest in the tender. Bad quality and slow delivery are the two most recurrent disservices received by the academic institution.

I have a colleague in one of the universities I work for who is directing another institute, with whom we have a running gag: "…and the cookies are disgusting!" We use it to conclude most conversations or complaints about bureaucracy at this

[3]By the way, I am not only mentioning Weber because he devoted a consistent portion of his work to management and bureaucracy. I am doing that also because he was *not* opposed to bureaucracy as such, but to its excesses only. One of the main points I am trying to make in this book is that the humanities are especially powerful in *critical* assessment of reality. Their power is exactly in not being monolithic, unilateral and therefore ultimately unfair. If *50 shades of grey* had not become the title for a silly book and an even sillier movie, we could have exactly said that humanists can turn a generic black-or-white analysis of reality into one that catches most of the grey nuances of it.

university. It refers to the fact that the tender for cookies, to employ in conference coffee breaks and the likes, was won by a company in a town that—note—is over 150 km distant from where the university is. They became the symbol of our dissatisfaction against the procurement system, because they are objectively and empirically of a very bad quality, with something red on top (sold as jam) which is probably silicon, judging from the consistency and heroic resistance to any attempt at chewing it. Now, everything is wrong about this procurement: (1) Imagine how cheap the production of such cookies must be, if the company managed to win the tender despite 150 km of distance (that is, the overall price we pay *includes* transportation, and still they manage to outdo local companies from the competition); (2) Imagine the contribution to environmental protection that a similar enterprise carries out: vans filled with horrible cookies travel 150 km to delight our congress guests, then return happily empty to the factory for another 150 km; (3) Imagine what progress in image and reputation our institution gains by inflicting inedible coffee breaks to our international colleagues, and what interesting academic discussions take place while half of the crowd is stoically working their tongue around to remove indestructible pieces of red silicon from their molars.

And in all this… the cookies are disgusting.

This colleague I mentioned once explained me how she got around this embarrassing obstacle. Take note: she pretended she had an important meeting in another city, so she applied for refunding to the faculty (being a very important city, this one is a very likely destination for one-day missions for all of us—so, usually, the financial department does not question too much the reasons behind such trip: a meeting with some colleague for one of the many projects we have sounds already very credible). We have a document with a table reporting fuel costs, related to the average consumption of various car engine types, so my colleague filled the tank of her own car with the equivalent amount needed for a trip back and forth to this city (200 km in total, for the record). Of course, you understand, it has to be a car trip, not a public transportation one, otherwise one actually needs to buy a ticket. Given her reimbursement, she finally had the money *not* to go to that city, but instead walk less than 100 m from the faculty, and buy delicious anti-procurement cookies from the bakery round the corner.

Another episode I may mention in the procurement odyssey concerns the hotels where we accommodate the invited speakers of the conferences we organize. One of the universities I work for arranged a tender to get special offers from three 4-star, three 3-star, and three 2-star hotels. The cheapest in each category would get the exclusivity for our accommodation requests. Mind you: the cheapest. That was the only requirement. Nobody thought it would be wiser to establish *a set* of parameters (prestige of the hotel, distance from the university buildings, absence of architectural barriers, attention for different lifestyles in their restaurant menus, nicety of the location, etc.) in order to choose the cheapest option that would satisfy all, or most, of them.

Results. Great for the 2-star winner: very nice and clean hotel, located on a beautiful avenue, and very close to the faculty: students and low-budget participants can be happy. So-so for the 3-star winner: very nice hotel, very nice location in a quiet neighbourhood, but also *very* far from the faculty: you need to use public transportation or a bicycle. Finally, terrible result for the 4-star winner, where of course we like to accommodate our plenary speakers, whom—as it happens—tend to be the most senior ones as well: the hotel as such is great, but it is not very close to the faculty—of that kind of distance that you might generally like to cover by walking, but not if you have your paper in less than 30 min, or if you have reached a certain age; and—most of all—the hotel is located in an area of the city, which is far from being the most attractive one. So, we give our most prestigious guests an image of ourselves as a city (and a country) that is not particularly appealing. And to add salt to the wound, this particular hotel won the tender by offering rooms that are only one euro cheaper than a hotel (which we would have loved to hire) located in front of the mentioned 2-star hotel—thus, very close to our faculty.

One euro.

One euro is enough to lose our face. This is what happens when these kind of calculations are made on an exclusively-quantitative basis.

Plus, of course, with only one euro of difference, you cannot help suspecting that a bit of behind-the-scene arrangement *did* take place, and someone, somewhere had some kind of interest to have that particular hotel winning the tender. But of course the procurement system is totally transparent, is it not?

2. Motivation to work for, and sense of belongingness to, an academic institution are discouraged and seriously endangered. Academics start perceiving the headquarters of their department/faculty/university as enemies who get in their way, do not trust them and merely harass them with unnecessary work. This is particularly a problem for humanists, because they are most often the "intellectual façade" of an academic institution. They represent it in the media with their magazine columns, TV appearances, interviews (a journalist who wants to write about a given university is more likely to approach the charmingly talkative philosopher from that university, rather than—say—the IT semi-autistic nerd[4]); they appear more often in public events, and so forth. An unmotivated, "unpatriotic" humanist, who is immediately visible and recognizable, is likely to be very vocal in his/her dissatisfaction, and is therefore the first to face the consequences.
3. The excessive reliance on schemes and regulations inhibit personal initiative and growth of the academics, who are treated more like machines than like individuals. The human factor is almost entirely neglected. For instance, current teaching programs, with all their requirements, precise scheduling, regular monitoring, fixed methodologies, "aims and objectives" mechanical templates, "at the end of the course students will be able to…" redundancy, etc., damage the teachers' abilities to accommodate their work to the students' need, and to assess them case by case. By now, teachers have to take too many *final* decisions *before* having even met their students. They are less and less enabled to understand what can be done, how, with whom, etc. They cannot "make a

[4]No offence meant. I hope it is clear I am talking about the stereotypical image projected by academics to the general public.

difference" as individuals, beyond institutional restrictions. Humanists pay the highest price for this, because they are naturally inclined to "make a difference" in their work by their personal talents, critical mind, individual style, unconventional choices and unique background. There is no need to recur to the fascinating scenario depicted by Peter Weir in *Dead Poets' Society*: a teacher can make the difference with a much less fancy approach than asking the students to step on their desks, read poems in a cave and call him "Oh captain, my captain". Still the point remains: a teacher can make students dig inside themselves and discover their true identity, showing them a path that will characterize their whole life. Teachers, in other words, have the ability to become mentors, if they are left free to do so.[5] Nowadays, mentorship tends to happen only as a result of a special relationship established between teacher and student *in their spare time*, that is, if they are able to develop that relationship outside the spaces and the times of their class.

4. And since we mentioned it, "time", in both the operative and the conceptual sense, get the most unfair treatment in the process. There is a fantastic website called phdcomics.com, where the most hilarious and effective jokes about academia can be found. One of such comics is entitled "How professors spend their time" (http://www.phdcomics.com/comics.php?f=1060), and emphasize the three main problems related to the way bureaucracy ("service") get in the way of an academic's job: (a) it concretely occupies a good quarter of the employee's working time; (b) it is not acknowledged as working time by the employer; (c) it is an extremely annoying imposition for the employee.

This is indeed the reality: bureaucracy takes a very consistent amount of time to be implemented, and yet does not get any recognition as "working time". Where do academics take time to, for instance, compile and re-compile ad infinitum their list of publications for different goals and destinations (e.g., library, doctoral committee, project department, human resources, mid-term report, full-term report, accreditation of their job, accreditation of their department, accreditation of their faculty, accreditation of their university, and so on and so forth), within a situation where not only the various units do not communicate with each other (they could share the same "list of publications" document, for a start, but they never do that, do they?), but they also have different templates, criteria, order, parameters for

[5]If a personal note is allowed at this point, I have to say that I know very well what I am talking about, as I owe my academic career (not only its contents and methods, but the choice itself to embark on one) to my two mentors, Gino Stefani and Eero Tarasti, who—each in his own way and style—understood things about me long before I myself would realize them. For instance, I will never forget how, in 1996, after a not very brilliant in-class presentation, Stefani told me and the whole class that with that "strength" I would have gone far in the academic world. This was particularly "avant-garde" not only because I did not feel I had displayed any strength on that particular occasion, but mostly because my decision to become an academic was only taken in 1999: in 1996 my dreams for the future were totally directed elsewhere.

compiling that single set of information. So, today my list of publications will have to be arranged in chronological order, Times New Roman 12 and MLA style; tomorrow it will have to be by levels (A first, than B, C...), Arial 11 and Chicago Manual of Style; etc.

All this takes t-i-m-e, but in the end of the year report, the employee cannot write that, say, 10 hours of his/her work were used to write and rewrite the list of publications for twenty different purposes, simply because there is no entry in the report template that says "time spent complying with administrative tasks". As hilariously-yet-tragically emphasized in the second statistic of the comic, that time is "outside", it is entirely neglected by the institution. Once more, humanists generally suffer the biggest handicap in this process, because their activities are also the most difficult (or at least varied) to categorize, and tend to be more numerous in quantity. If, for example, the activity of a biochemist is very focused on a single (very important) experiment for five years, and that experiment results in *one* (very important) article in a prestigious scientific journal displaying the success of that experiment and maybe a groundbreaking discovery, that of course (and justly) corresponds to a great achievement and a career high. The humanist, in that time span, was likely busy with 15 conferences, 10 articles, one monograph, 5 public lectures, 8 different (big or small) research projects, 3 edited collections, 2 book reviews, 10 Erasmus missions, the preparation of a new study program, and who knows how much more. All of this has to be reported, and very probably categorized in a different way than the one I just listed, which brings us to the next problem;

5. Along with a "time" issue, there is also a categorization issue. One reason why bureaucracy takes a lot of time is because it claims to be able to map all the variables within a given task, while it generally fails to do so. That is particularly true in the humanistic academic environment, where—generally speaking—things make sense when developed into a narrative structure, not into a "this or that" classification. The above-mentioned report, as compiled by a humanist, would make much more sense if the scholar was enabled to write it in free form, where s/he has the chance to show the consequentiality of certain activities (and related choices), as part of a perhaps eclectic but very coherent discourse. But no. At the entry "conferences" the scholar has to separate the international from the national; the ones related to a given research project from the ones related to another; the ones that will result in a proceedings publication from the ones that will not; the ones where the scholar was invited as key-note from the ones where s/he was not... At the entry "articles", s/he has to separate A level form other levels; peer-reviewed from not peer-reviewed; journals from edited collections; top publishers from less important ones...

The narrative is deconstructed and the result is a fragmented puzzle where the single pieces make much less sense than their sum. When this problem is raised, the hyper-bureaucratic apparatus responds in the worst possible way: it adds *more* variables, *more* categories, *more* tasks, in the vain illusion that sooner or later the full spectrum will be covered, and failing once again to understand that this is not (This. Is. Not.) a quantitative problem. When we were kids, we would realize pretty soon that with a box of six tempera colors we could not draw very imaginative pictures; so we would ask our parents to buy a 12-tube box, but that, too, would not be sufficient; so we would ask for a 24-tube box, until that would not be sufficient either… and so on and so forth, until we grow up a bit, and (click!) we realize that colors can be "mixed", and all we need is the three primary colors plus the two neutral ones. And from that small unit of five colors, we can develop any pictorial "narrative" we want, with all the tiniest nuances.

Needless to say, in this example hyper-bureaucracy corresponds to the stubborn child who keeps on asking for more and more colors, without realizing the qualitative aspect of those tubes: they can be mixed! This, too, takes us to the next step of the analysis;

6. There are two important paradoxes that result from this mentality, and they both have scientific names. One is the *Downs–Thomson paradox* about street traffic (Downs 1992: 30): if we try to solve the traffic problem by enlarging the roads and adding more lanes, we will only have more traffic. Translated: if we think that adding more procedures and more managing personnel (more tempera colors) will make bureaucracy lighter and more usable, let us think again!

The second one is known as the *Jevons paradox* (Jevons 1865: 146): an increase of efficiency generates an increase (rather than a decrease) of consumption. Translated: we are told that procedures are made more efficient (again, in a quantitative sense only: from 6 to 12 colors, from 12 to 24, etc.), but what happens is that we are getting more and more dependent on them. Bureaucracy develops in a rather organic way, and tends to fill all the empty spaces. Little by little, it gets to regulate not only the things that were in need of being regulated (if they really were), but also those that were simply standing in the way. What was before possible to do without, say, an authorization, now requires one; what was possible with two signatures now requires five, and so on.

Since we are into the realm of economic paradoxes and laws, perhaps we can also add a collateral effect of the constant reshuffling of personnel. It became known as the Peter Principle (Peter and Hull 2009: 9): anything that works will be used in progressively more challenging applications until it fails. It is a typical career-scheme in management (and a typical career goal of managers): if a manager X is good in the A position, s/he will get promoted to the B position (better paid, but with different tasks), if s/he succeeds there as well then there is C (better paid, but with different tasks), then D (better paid, but with different tasks)… until it turns out that, well, tasks became so different that X performs very poorly, the institution suffers a drawback, and X may also be fired (which, playing with words, may remind of how "fired" Icarus ended up for wanting to fly higher and higher). Of course, this situation tends to occur pretty soon (a promotion path from A to D and beyond is very optimistic, because normally X fails already at B or C), and activates a chain reaction: when X went, say, from B to C, manager Y was promoted from A to B (and maybe s/he will fail in this new position, while X was very good there), and manager Z was hired for the A position (where Y was very good). All this, while instead if someone is good at something, s/he should be kept there, and, if anything, given a higher salary and more bonuses for his/her merits. On the contrary, what happens (using football as an example) is that a good defender is promoted to striker (bigger salary and more fame), while of course s/he has no clue how to score goals.

To put these reflections in practice, we could for instance mention the data reported by the *Times Higher Education* (Jump 2015) about UK universities. There, we learn two important facts: that support staff has increased by no less than one third in only seven years (between 2003 and 2010), and that such staff is the majority at 111 out of 157 institutions. So, in more than 2/3 of British universities there are more managers than people to manage (with peaks of over 60 % in such cases as the University of Bradford, the University of Wolverhampton and Durham University[6]). Which is more or less (keeping up with the football example), like having a football team with 11 players on the pitch and 16 coaches in the dugout. At the same time (of course!) the best UK academic institutions (Oxford, Cambridge, Institute of Cancer Research, King's College…) are all in the Top 10 of those with *the lowest* amount of support staff (the lowest being ICR with only 37 %). So: where there is quality there is less (need for) management.

Maybe it is time to focus on quality, for a change?

The conclusion to this whole entry of the manifesto is therefore the following: in order to re-open the dialogue with institutions, we certainly need to *learn* the "rules of the game", and that is something we cannot escape from. At the same time, this must be counterbalanced by the equally-crucial necessity to *rewrite* such rules. In fact, we should learn such rules *because* we want to rewrite them, at least some of them.

Some of these rules have become necessary because of our laxity and selfishness: we need to admit that, and face the consequences. Some others must be instead understood in order to be corroded from the inside. The phenomenon of hyper-bureaucracy certainly belongs to the latter category. There is absolutely no doubt that every single scholar in this world, not just within the humanities, should roll up their sleeves (which is the point: "roll up their sleeves", not complain) to develop solutions against the outrageous overload of paperwork, evaluations, self-evaluations, monitoring processes, performance assessments, and all possible (imaginable and, more and more often, unimaginable) tasks the employees of academic institutions are systematically and increasingly harassed by. This is depressing, demotivating, frustrating and ridiculous. And counterproductive, for what is worth: the melancholic irony that these sadist[7] managers of knowledge are

[6]And I wonder what Mikhail Epstein thinks about such loads of management in the place where he has his Centre for Humanities Innovation.

[7]About sadism: not too long ago, during a faculty meeting, I sarcastically remarked that bureaucrats do actually enjoy making things more difficult for us. One colleague reproached me by saying that, well, expressing dissatisfaction for bureaucracy is one thing, but going as far as to say

obviously incapable of noticing is that most of these tasks are meant to "assess and improve" the quality of the academic work, but they are so time- and energy-consuming that the first and foremost process to get damaged by them is the production of quality itself.

"How do you work?", "How much did you work?", "Can you quantify this particular work?", "And what about that other work?", "How did you distribute the tasks of this work?", "Was this work performed within the S.M.A.R.T. framework?", "Can you do the S.W.O.T. analysis of your work?"... These and millions of others are the questions managers harass us by.

We, in turn, have only one question: "Can I go back to work, please?"

2.3 (Re)Learn the Noble Art of Empirical and Applied Approaches

Academic knowledge is traditionally divided into two main areas, the humanities and the natural sciences (though, by now, social sciences have become a credible third field, in between the previous two). Neatly separated nowadays, only a few centuries ago all sciences were branches of philosophy, which instead today is regarded as a sub-group of the humanities. As time went by, sciences grew more and more influential, autonomous and sophisticated, claiming space for separate labels; physics, chemistry, biology, but also sociology, economics (these two becoming eventually parts of social sciences), and so forth.

By 1959, the most famous acknowledgement of the existence of "two cultures" (separated by a "gulf of mutual incomprehension") took place, as the English physical chemist and novelist Charles Percy Snow delivered a talk in the Senate House, Cambridge, subsequently published as a book entitled *The Two Cultures and the Scientific Revolution*:

> I believe the intellectual life of the whole of western society is increasingly being split into two polar groups. When I say the intellectual life, I mean to include also a large part of our practical life, because I should be the last person to suggest the two can at the deepest level be distinguished. [...]. Two polar groups: at one pole we have the literary intellectuals, who incidentally while no one was looking took to referring to themselves as 'intellectuals' as though there were no others. [...] Literary intellectuals at one pole—at the other scientists, and as the most representative, the physical scientists. Between the two a gulf of mutual incomprehension—sometimes (particularly among the young) hostility and dislike, but

(Footnote 7 continued)

that bureaucrats are sadists is a bit too much. They, too, have to cope with rules, it is not easy for them either, etc. Well, that was of course a not-entirely-serious comment from my part. However, when you are shown—as I was—the document with the specifications for each of the faculty's procurements for coffee breaks, and you bump into the entry "apples", and you read that they have to be "second class" (in Lithuania there are three classes of quality for apples: premium, first and second—so we get the worst ones, but never mind) and (I kid you not) "not rotten yet", you know that somebody, somewhere, is having a lot of fun.

> most of all lack of understanding. They have a curious distorted image of each other. Their attitudes are so different that, even on the level of emotion, they can't find much common ground. Non-scientists tend to think of scientists as brash and boastful. They hear Mr. T. S. Eliot, who just for these illustrations we can take as an archetypal figure, saying about his attempts to revive verse-drama that we can hope for very little, but that he would feel content if he and his co-workers could prepare the ground for a new Kyd or a new Greene. That is the tone, restricted and constrained, with which literary intellectuals are at home: it is the subdued voice of their culture. Then they hear a much louder voice, that of another archetypal figure, [Ernest] Rutherford, trumpeting: "This is the heroic age of science! This is the Elizabethan age!" Many of us heard that, and a good many other statements beside which that was mild; and we weren't left in any doubt whom Rutherford was casting for the role of Shakespeare. What is hard for the literary intellectuals to understand, imaginatively or intellectually, is that he was absolutely right. (Snow 1959: 4–5)

A funny, little exercise (which, at the same time, is also a slightly melancholic one) that we can do to better understand this split, is to check on any encyclopedia how many great intellectuals of the past belonged to both humanities and natural sciences until the very early 19th century, and how many of them became only humanists afterwards:

- Francis Bacon, (22nd January 1561–9th April 1626): philosopher, statesman, scientist and jurist.
- Galileo Galilei (15th February 1564–8th January 1642): astronomer, physicist, engineer, philosopher and mathematician.
- René Descartes (31st March 1596–11th February 1650): philosopher, mathematician and scientist.
- Gottfried Wilhelm von Leibniz (1st July 1646–14th November 1716): polymath, philosopher and mathematician.
- Isaac Newton (25th December 1642–20th March 1726): natural philosopher, physicist and mathematician.
- David Hume (7th May 1711–25th August 1776): philosopher, historian and economist.

And so forth. As opposed to:

- Immanuel Kant (22nd April 1724–12th February 1804): philosopher.
- Georg Wilhelm Friedrich Hegel (27th August 1770–14th November 1831): philosopher.
- Arthur Schopenhauer (22nd February 1788–21st September 1860): philosopher.
- Ludwig Andreas von Feuerbach (28th July 1804–13th September 1872): philosopher and anthropologist.
- Martin Heidegger (26th September 1889–26th May 1976): philosopher.
- Jean-Paul Charles Aymard Sartre (21st June 1905–15th April 1980): philosopher, writer and literary critic.

Now. Despite this progressive separation, the humanities maintained for a long time a clear centrality within intellectual inquiry, via the commonly-acknowledged assumption that the liberal arts are at the basis of personal realization, education and citizenship (and perhaps—as Snow had suggested—also because humanists were very quick in taking possession of the social label of "intellectuals"). Little by little (and—as one may have guessed from the previous list of philosophers pre- and post-19th century, I personally locate this process in a period between Romanticism and Postmodernism, therefore in the last two centuries), that centrality turned into a marginality (nowadays bordering with irrelevance) as it became clear that humanists were detaching themselves from reality, current concerns, socially-relevant matters, and empirical evidence, while the sciences were becoming more and more able to go the opposite way, and even be increasingly understood in the process (with "popularizing science" media programs and writings, with scientists like Einstein becoming pop icons, etc.).

By the 21st century, the humanities do not seem to be anymore able to "speak to the world" with the language of constructiveness and concreteness. Elaborating a bit on what Sokal (2008: 229) says specifically about postmodernism (on which we shall return in point 7 of this manifesto), we could say that this detachment materialized in two ways, "one crude, the other subtle" (as Sokal himself has it). The crude detachment was an opposition between objectivism/realism/pragmatism (represented by sciences) and relativism/subjectivism (as proposed by humanists). The subtle detachment was an opposition between that same realism and, on humanities' part, a form of anti-realism/instrumentalism. In the former case:

> Relativism and radical social constructivism have become hegemonic in vast areas of the humanities [...]. In many intellectual circles nowadays, it is simply taken for granted that all facts are "socially constructed", scientific theories are mere "myths" or "narrations", scientific debates are resolved by "rhetoric" and "enlisting allies", and truth is a synonym for intersubjective agreement. [...] in numerous debates with sociologists, anthropologists, psychologists, psychoanalysts and philosophers, [...] we have met people who think that assertions of fact about the natural world can be true "in our culture" and yet be false in some other culture. We have met people who systematically confuse facts and values, truths and beliefs, the world and our knowledge of it. Moreover, when challenged, they will consistently deny that such distinctions make sense. Some will claim that witches are as real as atoms, or pretend to have no idea whether the Earth is flat, blood circulates or the Crusades really took place. Note that these people are otherwise reasonable researchers or university professors. All this indicates the existence of a radically relativist academic Zeitgeist. (Sokal 2008: 229–231)

As for the latter case:

> Roughly speaking, realism holds that the goal of science is to find out how the world really is, while instrumentalism holds that this goal is an illusion and that science should aim at empirical adequacy. [...] Relativists sometimes tend to fall back on instrumentalist positions when challenged, but in reality there is a profound difference between the two attitudes. Instrumentalists may want to claim either that we have no way of knowing whether "unobservable" theoretical entities really exist, or that their meaning is defined solely through measurable quantities; but this does not imply that they regard such entities as "subjective" in the sense that their meaning would be significantly influenced by extra-scientific factors (such as the personality of the individual scientist or the social characteristics of the group to which she belongs). Indeed, instrumentalists may regard our scientific theories as, quite simply, the most satisfactory way that the human mind, with its inherent biological limitations, is capable of understanding the world. (Sokal 2008: 231–232)

In order to approach the problem in a constructive way, we need to answer a few questions: What does it mean to be concrete and empirical, when it comes to the humanities? Can this be still achieved nowadays, or was it just a result of an outdated universalistic/encyclopedic conception of knowledge production? And most of all, is it worthwhile?

The thesis we want to defend, within the project of Numanities, is that the humanities have both the necessity and the potential to recover (and update/upgrade) that pre-Romantic inclination to combine what the American Educational Research Association calls "conceptual inquiry" with empirically-based scholarship. Although there is no pretension here to provide an exhaustive overview of the many existing attempts to integrate these two approaches (I shall limit myself to very few cases, plus some personal reflection), we may at least start by underlining that

> the humanities and the sciences are not so disparate as many have been led to believe. Although some have argued that numbers simply aren't important or appropriate in the humanities, if a literary critic were to argue for a new interpretation of why Hamlet sent Ophelia to the nunnery and gave only one line from the play as evidence, the critic would be laughed at by responsible literary scholars. Thus, numbers are important to literary scholars. In a like vein, music is generally regarded as a purely aesthetic experience, totally unrelated to science, yet a melody can be described very precisely with numbers. In fact, equipment that plays music which has been recorded digitally (i.e., numerically) is reputed to be the equipment of choice for serious listeners who want the highest quality of reproduction. Finally, even philosophy owes much of its development and theory to numbers witnessed by the important role zero plays in logic. (MacNealy 1999: 5)

The similarity becomes more tangible when MacNealy (1999: 7) makes a practical comparison of the methodological paths of the "two cultures", showing that, at the end of the day, similarities are greater in number than differences:

Library-based research (typical of the humanities)	**Empirical research** (typical of natural sciences)
1. Encounter a dissonance	1. Encounter a dissonance
2. Define the problem	2. Define the problem
3. Define the research question(s)	3. Define the research question(s)
4. Plan search of published ideas and observations of others	4. Plan for systematic collection of directly observed activities, products, and other phenomena
5. Search for and analyze evidence—Methods: deductive, analogical	5. Collect and analyze the data—Methods: descriptive, experimental
6. Propose new theory or insights in a thesis and support essay format	6. Interpret findings in relation to theory and prior findings in an IMRAD format (introduction, methods, results, and discussion)
7. Argue for acceptance and significance	7. Argue for acceptance and significance
Purpose: Build theory Result: Theoretical (provide hypotheses)	Purpose: Build theory Result: Descriptive (provide or test hypotheses), Experimental (test hypotheses)

Also, MacNealy (1999: 8–10) argues that "lore" (accumulated knowledge, cultural experience, etc.) is an added value for both areas, serving often as a background and facilitator for research. Assuming thus lore as a special "eight step" (a bit like George Martin served as a fifth Beatle), we notice that five steps out of eight are identical, plus the purpose, while three steps and the category "result" differ. Moreover, the similarities show that, generally speaking, humanities and sciences tend to start and conclude inquiry in the same way and with the same purpose. All considered, this should not be really surprising, as it is only natural that all knowledge processes depart from the same point and try to reach the same destination. What changes significantly, besides the result, is the definition of what data are, their management and their interpretation (to make an obvious example, humanists tend to identify data collection with the search and analysis of other scholars' already-published ideas and reflections, while scientists tend to prefer direct observation and measurements).

A source suggesting that, in principle, also the treatment of data could be done in the "scientific" way is the very interesting Peer et al. (2007), which not only makes a solid point on the importance of adopting empirical research in the humanities, but it also embarks into a structured categorization of the most important and effective forms of empirical inquiry that are suitable for the humanities. These are:

- The various types of research methodology
- The various types of research design
- The methods for data collection (with a particular focus on questionnaires)
- The experiment as such
- The various forms of statistical analysis (with a particular focus on SPSS software, descriptive statistics and inference statistics)
- The communication of results

There is of course no need to go into the details of these suggestions, but what interests us is that they show, and convincingly argue about, the existence of a concrete possibility for the humanities to approach data in a fully empirical way. However: now that we understand that this is *possible*, the main question is: is this *needed*? I am definitely not posing the question in terms of "identity of the humanities": in fact, my humble suggestion is that we should not be too concerned with identity/authenticity problems at all, but rather with the question of making the humanities a more effective and credible form of knowledge production. Plus, if we really need to address ethical issues, I would rather insist on the need to recover dignity, rather than preserve identity.

Having said that, if identity remains a problem for some colleagues, I would not see why those particular stages of the research processes should be problematic. And I say this with the same emphasis I employ in maintaining that reading an e-book instead of a printed book does not pose any threat to our identity as readers. Not a single comma of the book we are reading changes, and even our beloved image of coziness consisting of reading a nice book on a sofa, by the fireplace, with a nice cup of herbal tea, while outside it is raining, cannot be really affected if what we hold in our hands is a tablet, rather than an object of the same size and weight, except that it is made of paper.

Identity is something that we must search for in the typology of knowledge we produce, not in the modes of production. And, after more than a century, it is difficult to find a better definition of such typology (as opposed—one in fact complementary —to the typology pertaining to natural sciences) than the one we find in the hermeneutics of Wilhelm Dilthey. Dilthey regarded the cognitive operations of *understanding* and *explaining* as two opposing modes of coming to terms with the reality that surrounds us. Explaining ("erklären") is what characterizes the natural sciences: in them, causality is the central concept; a scientist is interested in explaining how the A element/phenomenon causes the B element/phenomenon (e.g., how a given virus causes a given disease). Understanding ("verstehen"), on the other hand, is what characterizes the humanities (and, I add, what the humanities do best, and better than other fields). In this, interpretation, rather than causality, is the central concept: a humanist is interested in discerning and recognizing the individual-psychological and social-historical conditions that made it the B element/ phenomenon (which, as natural sciences have told us already, was caused by the A element/phenomenon).

> That which has developed in the course of human history and which common usage has designated as "the science of man, of history, and of society" constitutes a sphere of mental facts which we seek not to master but primarily to comprehend. The empirical method requires that we establish the value of the particular procedures necessary for inquiry on the basis of the subject matter of the human sciences and in a historical-critical manner. The nature of knowledge in the human sciences must be explicated by observing the full course of human development. (Dilthey 1991: 57)

To keep up with the same example as before, while sciences have given us the most precious information that it was *that* virus to cause *that* disease (plus additional important information on how it was formed, what was its chemical composition, what could be the cure, etc.), the humanities complement that part of the story by analyzing the conditions and impacts of the disease (e.g., it is the result of a given psychological/existential condition of the subject, it affects certain layers of society, it generates forms of social discrimination, it is not properly faced by institutions, etc.), adding *significance* to *meaning*. There are also cases (AIDS being possibly the best-known) where the "significance" provided by the humanities ended up creating more awareness and a more effective response than the "meaning" produced by natural sciences (exposure of homophobia, marginalization of the ill, religious prejudice… not to mention the crucial role played by popular culture characters and events).

At this point, it becomes clear why the humanities would greatly benefit from incorporating a bit of the scientific method in their inquiries. Providing "understanding" instead of "explanation" is an important responsibility, to be kept as much "feet on the ground" than the natural sciences' counterpart—unless, once again, one wants to lose credibility as "science". If, among the interpretations of a given disease, a humanist started including propositions like "divine punishment", "consequence of a black cat crossing a street", or "result of Venus's conjunction with Mars in Aries", then we know we are in trouble: with science, with institutions and (if we are honest) with ourselves. Even worse, to some extent, would be the waste of time of radical relativism and postmodern deconstruction of the problem: "Depends how we define 'virus'", "diseases are hyper-realistic simulacra", etc.

Although a religious man and an intellectual product of Romanticism, Dilthey was an empiricist. His work on the philosophy of humanities was never detached from a scientific view of reality. He did not consider "humanities" as something different from "science", but just as a different (complementary) kind of science than the natural ones. His encompassing definition of "science" is something we should always keep in mind:

> By a "science" we commonly mean a complex of propositions (1) whose elements are concepts that are completely defined, i.e., permanently and universally valid within the overall logical system, (2) whose connections are well grounded, and (3) in which finally the parts are connected into a whole for the purpose of communication. The latter makes it possible either to conceive a segment of reality in its entirety through this connection of propositions or to regulate a province of human activity by means of it. The term "science" is here used to designate any complex of mental facts which bears the above characteristics and which therefore would normally be accorded the name "science". (Dilthey 1991: 56–57)

Losing contact with reality is something that the humanities should seriously commit to avoiding, in order to recover dignity and credibility, and in fact—looking at their original status—identity too. Once more, in our attempt to analyze the situation in self-critical terms: there is a chance that the crisis of the humanities is not only due to the wild capitalistic mentality of most modern societies (an undeniable factor, of course), but also to those many humanists who got stuck somewhere between the romantic anti-rationalist individualism and the postmodern self-indulgent disillusionment of their ability to actually be *rational and concrete*.

While this particular point of the manifesto is not a plea against metaphysics, hardcore relativism or speculation per se, one must be aware that there is a risk in activating—say—a philosophical process that has no other goal than being speculative.

Among the much bad news involving humanities in this epoch, one is certainly positive: those institutions that are still promoting humanities in their programs have understood that a major strength of this field is to contribute to creating a "value-based society". As we have seen, that keeps on happening also when humanities are joined by social sciences.

The philosopher (or any other humanist) who is ready to engage in metaphysical or ultra-relativist thinking (and, at the same time, to complain that no funding institution will understand the importance of it), might perhaps benefit from starting with the following questions (not to mention more empirical ones): does this process create *a value* (and, make no mistake, I am not necessarily talking about an *economic* one)? Do I aim to apply my precious reflections somewhere, to some existing problem of society? Will I improve somebody's life? Am I speaking to the world (as opposed to "myself")? If all questions are answered by "no", the philosopher still has the right to pursue his/her intuitions, but also has a duty not to be surprised if nobody finds motivations to support an initiative that is so clearly private. The great American composer Charles Ives always refused to make a job out of his art for this very reason: he wanted to retain the rights and not bother about the duties.

At the same time, the fact remains that most great humanists of the past, in all fields, had the gift of connecting great visions with great actions, or—in the most forward-thinking cases—they *set the bases* for their vision to become *real* eventually (and by vision I do not only mean reflections, but actual projects and plans: Leonardo's machines were incredibly *avant-garde* and visionary, and yet perfectly rational and concrete).

George Doran's S.M.A.R.T. criteria are nowadays a fashionable way to set objectives, in projects (sometimes even *too* fashionable). A project for rational humanities has certainly to be SMART: it has to be Specific (by targeting a specific area for development), Measurable (which indeed means empirical), Assignable (and—in comparison to Doran's formula—I would not only refer to human resources, here), Realistic (feet on the ground, indeed!), and Time-related (an idea of applicability has to be foreseen at some point).

However, if we want our own conceptual acronym (and—as you will discover during this book—we love both acronyms and conceptuality!), I would suggest that we keep it R.E.A.L. (see also Fig. 2.1):

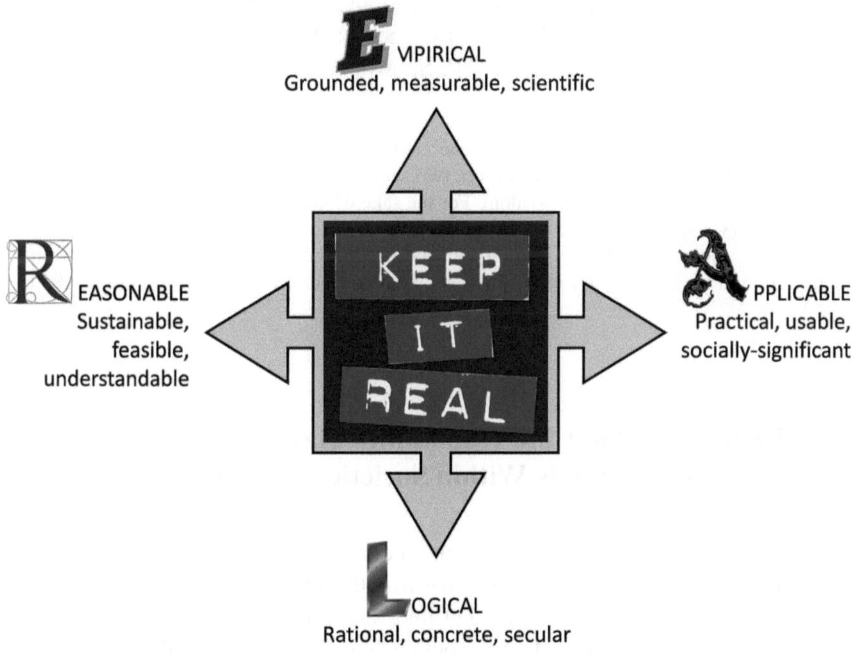

Fig. 2.1 The "Keep it R.E.A.L." principle for the humanities

1. Reasonable: sustainable, feasible and understandable. As much as possible, we should keep our research objectives within a circumscribed field, and our argumentations into a mode that limits the excesses of relativism, subjectivism and the likes.
2. Empirical: grounded, measurable and scientific. As much as possible, humanistic inquiry should develop arguments that can be demonstrated and validated. It is not necessary to always design and perform a laboratory experiment (unless the type of research calls for it), but it is important to employ tools that bring those arguments outside the purely abstract/speculative realm.
3. Applicable: practical, usable and socially-significant. As much as possible, we need less speculation for speculation's sake, and more understanding of the current world, more creation of values, and—why not—more applied humanities.
4. Logical: rational, concrete and secular. As much as possible, the exaggerations of metaphysics, anti-realism and the likes should give way to tangible and scientifically-plausible forms of research. As humanists, we need to speak again to the world, by remaining *in* it.

I have been using the expression "As much as possible" in all the four entries of the acronym, because I intend to point out that a humanistic inquiry that meets all the four requirements of the formula "at full speed" exists only in an ideal condition, and it may not be very likely in reality. What is important is to orient humanistic research in these four directions, and try to cover enough ground to make this research scientifically and institutionally credible and dignified, and to re-form a powerful unit with the natural sciences that can be very beneficial for both sides and, most of all, for society. C.P. Snow's conclusions to his lecture are still resounding today:

> Closing the gap between our cultures is a necessity in the most abstract intellectual sense, as well as in the most practical. When those two senses have grown apart, then no society is going to be able to think with wisdom. For the sake of the intellectual life, for the sake of this country's special danger, for the sake of the western society living precariously rich among the poor, for the sake of the poor who needn't be poor if there is intelligence in the world, it is obligatory for us [...] to look at our education with fresh eyes. (Snow 1959: 53–54)

2.4 Remember That the Humanities Have Always Taken Progressive Stands Within Societies, Not Luddite Ones

The humanist, the philosopher, the artist, the architect... for centuries, these roles were played by people who *looked forward* (the literal translation of the French *avant-garde*), who were able to predict mid- and long-term orientations in history (at various levels: creative, ethical, sociological...) by means of what, after the previous point, we may call "pragmatic speculation": a form of reasoning that derives from a deep understanding of the mechanisms and actions of a society, both at small and large scales.

There is a risk, for humanists, to become *uncritical scholars of the past*, and occupy the scholarly niches of traditions, folklore and philology, with an exclusive interest in memory, conservation and identity. While these, in themselves, are enterprises of enormous importance (there is no need to explain why), the crucial contact with *values* is occasionally endangered: an attitude of the type "traditions must be preserved because they are traditions" turns a good opportunity for reflecting upon a community's identity and *its future*, into a tautological nostalgia for the "good old days". When researchers do that, they are simply not doing their job properly, allowing personal biases to creep into what, instead, should be a solid scientific attitude. Failing to keep up-to-date with a society that is clearly not placing the humanities at the center of attention, may resolve into pub-like reminiscences of the past, rather than a proper study of it. Once again, the bound with reality is lost.

On the contrary, the role of the humanities, when dealing with the "past" (or any other topic), is to approach it in its organic, vibrant nature (e.g., as a key to interpreting the present or even predicting the future), never losing touch with reality, simply because the essence itself of the humanities is to provide the right (and the deepest) tools to interpret it.

Digital technologies are one of the best examples of how, generally speaking, technophobic humanists are missing an opportunity to connect the dots across the past, present and future. The sight of an e-book causes a melancholic preoccupation on how people are not reading *real* books anymore, how libraries are emptier, "where will we end up?", and so forth. But what is a *real book*? Is it by any chance that printed object that before Gutenberg was *not* a real book? A tradition is always preceded by another tradition: printed books are less traditional than hand-written manuscripts, which are less traditional than stone carving, which are less traditional than inarticulate vocalizations.

The risk is that there is not really a defense of tradition, or integrity, or identity, but rather of *familiarity*. Maybe we defend what *we know best*, not what is best in principle. The "unknown" makes us fearful and lazy.

I shall devote one of my case studies to the topic of authenticity in music, emphasizing—in its conclusions—how the ideals of originality, truth, tradition and others encourage the creation of a transcendental discourse category that fails to define itself in spatial-temporal terms and become an ambiguous meta-dimension. What and where are "the good old days"? At which exact point of (human or even pre-human) evolution shall we reasonably locate the famous tradition to preserve and re-install?

In her very interesting book on Luddite and neo-Luddite movements, Fox (2002: 330–365) takes the recent example of a magazine called *Plain*, printed by a community of Amish people in Ohio. Its editor in chief, named Scott Savage, makes a very firm point on the magazine's opposition to a modern life that is "moving at electronic speed. We're finding that people can't live at that speed. We're being crushed by that speed" (quoted in Fox 2002: 330). *Plain* has a specific editorial and lifestyle policy, articulate in several rules, and actively promoted among its readers. The first issue had the following statement of purposes:

> Our mission is to develop a new fellowship of people who want to make peace with nature, create viable communities and strong families, and lead spiritually based lives. [...] We seek to open a space where people enmeshed in contemporary culture can begin to discuss its effects on their daily lives. A space where encounters with simpler, less mediated living can be shared. Where the unity of life, land and spirit can be experienced. (quoted in Fox 2002: 331–332)

And, Fox herself adds, although Savage is a deeply religious man,

> he was starting from a different perspective than the Amish who, as the world changed around them, refused to accept those elements that interfered with their religious focus. Savage was rejecting a world that was already here because it was interfering not only with religion but with every other aspect of human life. He was seeking to unify what had become divided and isolated, to bring people back to older values as exemplified by older ways. He had sampled the new world and was intentionally turning his back on it. (Fox 2002: 332)

Among other rules, *Plain* objects to having any reference to itself on the internet, it does not have a phone line but communicates only via airmail, it is printed with the letterpress method, it features woodcut illustrations, and of course when Fox and Savage agreed to meet in New York for an interview, the latter preferred an overnight journey by train, rather than a practical flight.

Now, it is absolutely not my intent to criticize *Plain*, or similar enterprises. As a matter of fact, I have huge respect for anybody who goes all the way into the pursuit of an ideal, especially if this is an ideal for a better world. I am merely using a well-documented case (in a well-documented book like Fox 2002) to make my point. What is it, exactly, that Mr. Savage is against, and what is it that he is defending? Most of all, *which epoch* (or stage of human evolution) is he defending? The internet was invented in late 20th century, letterpress printing in the early 15th, xylography in the 3rd, rail transport in the 19th. That is: if we are not counting all forerunners, early experiments, geographical variations and the likes (e.g.: the World Wide Web was invented in 1991, but the precursor ARPANET is from 1969). All of these inventions were initially met with resistance from communities that were accustomed to previous technologies, but—what is much more important—they were *all* inventions. They were *all* technologies. A train is not less of an invention than an airplane. The resistance to railways generated in the 19th century by Luddism was something rhetorically expressed with the very same spirit as Savage's concern about the present. I am absolutely sure that, after the invention of trains, plenty of people must have expressed themselves with those same words: "We're finding that people can't live at that speed. We're being crushed by that speed". How can Mr. Savage travel on a means of transportation that was the very symbol of the Industrial Revolution that Luddites were so much against? Why something that was the embodiment of dehumanization is now the embodiment of those-were-the-days coziness?

These are of course rhetorical questions, as the answers are plain to see: historical conditions change and must be contextualized with all their various contingent variables. Such contextualization is not an easy task, because it requires a degree of detachment from our individual condition, which may or may not be compatible with the social picture and the demands generated by the technological advancements (needs for new competences, to begin with). This, as very well argued by Brosnan (1998: 60–78), provokes anxiety, fear and lack of self-confidence in the subject.

There is nothing particularly special in *this* specific era, in terms of technologies and progress. The impact that the internet, cloning or touchpad screens have had on society is not of a different kind than that of the wheel, electricity or firearms in their respective times. On the other hand, one major difference may exactly lie in the attitude taken by humanists in the process. Generally progressive and forward-looking before Romanticism, artists and intellectuals began more and more replacing the concept of "utopia" with "dystopia" in their perception of the future, as in this famous passage by Thomas Carlyle:

> Our old modes of exertion are all discredited, and thrown aside. On every hand, the living artisan is driven from his workshop, to make room for a speedier, inanimate one. The shuttle drops from the fingers of the weaver, and falls into iron fingers that ply it faster. The sailor furls his sail, and lays down his oar; and bids a strong, unwearied servant, on vaporous wings, bear him through the waters. Men have crossed oceans by steam; the Birmingham Fire-king has visited the fabulous East; and the genius of the Cape, were there any Cameons now to sing it, has again been alarmed, and with far stranger thunders than Gamas. There is no end to machinery. Even the horse is stripped of his harness, and finds a fleet firehorse yoked in his stead. Nay, we have an artist that hatches chickens by steam; the very brood-hen is to be superseded! For all earthly, and for some unearthly purposes, we have machines and mechanic furtherances; for mincing our cabbages; for casting us into magnetic sleep. We remove mountains, and make seas our smooth highway; nothing can resist us. We war with rude Nature; and, by our resistless engines, come off always victorious, and loaded with spoils. (Carlyle 1858: 13)

As popular culture came onto the scene, dystopia and technophobia became even more evident presences (see the very interesting Jones 2006: 173–210). For instance, by the 1970s already it had arguably become the most recurrent rhetorical device employed in popular music (certainly within the area we commonly identify as "rock") to address one's own dissatisfaction towards a given social-political situation, or—better said—"the system". The myth of "the system" as an evil, conspiring force that is distant and in fact opposed to "the people" (in turn a counter-myth) is a true topos that we find in literally hundreds of songs. Future hyper-technological, de-humanized, post-atomic, mind-controlling, Langesque-Kubrickesque depictions pop up so often that the temptation is to conclude that George Orwell's *1984* is by far the most-read literary work by rock stars. Examples include songs like Pink Floyd's "Welcome to the Machine" (where "the system" appears in the ever-hated disguise of the music industry), Emerson Lake and Palmer's "Karn Evil 9" (on the album *Brain Salad Surgery*, featuring a cover by H. R. Giger, a dystopian painter by definition), or entire albums like Frank Zappa's

Joe's Garage or David Bowie's *Diamond Dogs* (directly inspired by Orwell's masterpiece).

In other words, while always the result of a natural psychological process of fear and anxiety towards change, technophobia managed also to become a genre in humanities, or perhaps even a narrateme within their discourse. Once again, a big opportunity is lost (or at least, weakened): the humanities, methodologically, could be the most powerful and most reliable source for a critical assessment of technological advancement. Rather than withdrawing their hands from touching them, they could well dig into technologies in order to draw a credible interpretation of their state of the art, their potentials, their limitations, their advantages and their dangers. They could tell us much more about the difference between a tablet application that fights social discrimination and one that instead favors it, because they could work on subtleties that neither natural nor social sciences would be able to grasp (see also point 6 of this manifesto).

More to the point, they could offer another angle than just conservative resistance to new technologies. Interestingly enough, the mentioned *Plain* magazine seems to be particularly fond of the concept of appropriate technologies (of which, as I shall later declare, Numanities are fond as well, though in a re-adapted form). Ernst Schumacher is, for instance, often quoted by Scott Savage in his articles (Fox 2002: 332). But even admitting that the great German economist was skeptical of technologies in principle (and I myself have a rather different interpretation of his writings), how can one conciliate such feeling with the plain evidence that several new technologies are much more environmentally-friendly than old ones (and sustainability is an absolute pillar of the concept of appropriate technologies)? Can one really dare to compare the ecological impact of millions of printed copies of, some 700-pages-and-counting, Dan Brown's novel with the electronic version of it? Is a tablet not a great embodiment of the "small is beautiful" principle when it allows me to carry a whole library in a 20 × 30 cm device? Are those miraculous prosthetic limbs of nowadays not "appropriate" for a given PWD, just because they come from the latest technology? Is having defeated tuberculosis inappropriate?

Common sense suggests that, between technophilia and technophobia, one should find a golden middle. Nothing could be truer than that. But this middle should not be measured with a ruler: it should be a dynamic entity that may change its exact location and size, depending on each single case. The humanities could be exactly the added value that identify and describe these specific locations. Back to my different interpretation of the concept of "appropriate technologies" (on which, I repeat, I shall return more extensively), perhaps I can mention here that I see that concept *exactly* as a successful attempt to find a few of those many golden middles we need. I totally agree with Steven Jones when he says that, far, from being technophobes, the supporters of appropriate technologies "were pro-technology, as long as it was appropriate technology. It is sometimes forgotten that some counterculturalists were avid technophiles who mixed do-it-yourself pragmatism with radical optimism and shared a vision of technology's utopian potential for building a new kind of community" (Jones 2006: 181). This is the type of counterculture that brought straight into the creation of "open source" principles in software and operating systems, which is *the* appropriate technology by definition in the IT environment, despite whatever any Apple fanatic will tell you.

In life, the turning point between adult and senile age is not an anagraphic one, but occurs at the very moment when the enthusiasm for learning and exploring is replaced by the illusion of having found "the truth" in the already-learned and the already-explored. This is why some of the most prestigious academic institutions in the world (King's College, Harvard University, MIT…) have embraced excellent research and teaching programs in Digital Humanities: they are consciously refusing to "get old".

The point in Numanities is not to accept new technologies in principle (we shall see this in the next point), but rather to give proper, careful scrutiny to them: scientific criticism, when needed, should be grounded and motivated by scientific, not nostalgic, claims. The reflections should be produced in a socially-contextualized manner, not out of individual urges or frustrations.

2.5 Be Ethically-Minded, Sustainable, Oriented Towards Quality and Dignity of Life

In the words of Gandhi first and Schumacher later, progress is implemented through *appropriate technologies*. "Progress" is not necessarily represented by every single thing that is newer, bigger, faster: a critical (therefore, intimately humanistic) view seeks progress in what is tailored for the real needs and demands of a given community (indeed: what is "appropriate", and that—as we have learned—may also mean existing, local, slow technologies).

Humanities have the theoretical skills to offer a constructive and critical view on these matters: Numanities would like to re-establish this interest and additionally cultivate operative and concrete tools not to let those skills be wasted in purely abstract, distant-from-reality formulations.

To formulate the question in a catchy manner, the point about progress is *not* about going forwards or backwards. The point is to go *towards*. Towards the people, towards the society, towards the environment, towards life.

If we take, as example, Daniel Raphael's recent Theory of Human Motivation (Rapahel 2015), we understand why it is necessary to talk about *both* quality and dignity of life. "Dignity" is a concept that expresses an intrinsic right to be (a) valued and (b) treated ethically. Raphael discusses three basic values, as foundations of human motivation: quality of life, growth and equality. Quality of life does not presuppose "equality" (it does not oppose it either, of course), dignity of life does. Quality of life pursues growth for a "liveable society", dignity of life pursues it for a "just society". Quality of life wants citizens to be treated well, dignity of life wants them to be treated ethically: the member of a given ethnic minority who is wealthy beyond his/her basic needs and is properly educated and assisted by the state, is a subject who receives quality of life; but if, in the meanwhile, s/he is still discriminated against for belonging to that ethnic group, then s/he is not given "dignity of life". Still continuing with Raphael, I see quality of life more overlapping with "material sustainability" and dignity of life with "social sustainability". Finally, quality of life produces "well-being", while dignity of life produces "values".

The concept of appropriate technology (AT)—sometime known also as "intermediate technology", particularly via Ernst "Fritz" Schumacher's work (of whom we shall soon talk about)—originates from Gandhi's struggle for Indian independence, when he developed the idea of economic "self-reliance" (in Hindi "Swadeshi"), that is the advocacy of small and local technologies by which Indian villages could reject economic colonialism from Britain. Embodied by the famous image (and flag symbol) of the spinning wheel, swadeshi produced the khadi movement in the 1920s, that is the local production of clothes as a form of civil disobedience, eventually causing the collapse of the British monopoly on textiles.

Starting from the 1950s, the German economist Ernst Schumacher, inspired by these ideas of Gandhi, brought the concept of AT to a more systematic level, particularly via his famous book *Small is Beautiful* (1973). What emerges from Schumacher's work, and that kind of became the three pillars of the whole theory, is the following three characteristics of AT:

1. AT's have to be **sustainable**: that is, environmentally sound and "economic" in the literary sense of the word. Not by coincidence, the concept of AT, which nowadays is certainly less fashionable than in the 1970s, has been partly replaced by "sustainable development": they are not the same things, but plenty of what sustainable development theories advocate were originally promoted by AT theories.
2. AT's have to be **appropriate** to a given context. Contexts are of an ethical, cultural, social, economic, environmental and political type. Obviously, a technology that is appropriate for a given context, is not necessarily so for another, however some generally-applicable rules exist: community input, affordability, reparability and maintainability are certainly four universally "appropriate" properties for a technology
3. AT's have to be **small** where possible (or, as I prefer to say, "direct"). Of the three pillars, this is the one that is most often misunderstood, since "small" is naturally perceived as unambitious, unpretentious, modest. On the contrary, the main idea is to place more power at the basis of an economic process, at the grassroots, and to make sure that the technology is in the hands of the users. In that sense, I prefer the word "direct", in the same sense that one would employ in speaking about "direct democracy". Also, the idea of appropriateness itself implies proportion and context-dependence. Technologies can therefore be also "big", if the context requires it:

> The question of scale might be put in another way: what is needed in all these matters is to discriminate, to get things sorted out. For every activity there is a certain appropriate scale,

> and the more active and intimate the activity, the smaller the number of People that can take part, the greater is the number of such relationship arrangements that need to be established. Take teaching: one listens to all sorts of extraordinary debates about the superiority of the teaching machine over some other forms of teaching. Well, let us discriminate: what are we trying to teach? It then becomes immediately apparent that certain things can only be taught in a very intimate circle, whereas other things can obviously be taught en masse, via the air, via television, via teaching machines, and so on.
>
> What scale is appropriate? It depends on what we are trying to do. The question of scale is extremely crucial today, in political, social and economic affairs just as in almost everything else. What, for instance, is the appropriate size of a city? And also, one might ask, what is the appropriate size of a country? Now these are serious and difficult questions. It is not possible to programme a computer and get the answer. (Schumacher 1973: 50)

Still during the 1970s, another important AT scholar, the Austrian designer Papanek (1971), added another piece to the theory by emphasizing the concept of "design", and specifying that the design of an AT has to always keep in mind two goals: the "real world" and the "human scale", both often overlooked in the diverse economic processes. As the definition of "real world" is probably quite obvious, the following quote from Papanek's book may perhaps better clarify what he meant by "human scale":

> The whole concept of human scale has gone awry, not only with homes but in most other areas as well. One would expect a system motivated only by self-interest and private profit-making at least to spend some care in constructing its shopping places. This is not so. 'Stroget' a 'walking street' of shops in downtown Copenhagen, is constructed for leisurely strolling and impulse buying. Two segments of it, Frederiksberggade and Mygade, are together approximately 400 feet long and contain more than 180 shops.
>
> In a contemporary American shopping centre, this same distance of 400 feet will frequently separate the entrances of two stores: the supermarket and, say, the drugstore. The intervening space consists of empty windows, bereft of displays, monotonous and uninteresting. Usually neither landscaping nor wind-breaks are provided. Mercilessly, the hot sun beats down on the four acres of concrete in the summer; wind-whipped snow piles up in car-high drifts throughout the winter. Small wonder that, after finishing their shopping at the supermarket, people will walk back to their automobiles and drive to the drugstore. There is nothing in the environment that prompts going for a stroll; it has been designed for the moving car alone. Most shopping plazas in the United States consist of a thin line of stores arranged along three sides of a huge square, the centre of which is a parking lot. The large open side fronts on a super-highway. This may make shopping 'efficient' but it also makes it something less than satisfying. (Papanek 1971: 258–259)

As—predictably—the main point of this manifesto entry is to make an economic concept "function" within the abstract level of an academic-scientific context like humanities, it shall be also important to mention Audrey Faulkner's and Maurice Albertson's distinction between hard and soft AT's (1986). While hard technologies are grounded in the natural sciences, scientific techniques, and improve upon tangible structures, soft technologies concern the needs and values of a community, and are grounded on social structures, interactive processes and motivation techniques. My conviction is that an area of academic/scientific investigation can be configured for society in form of soft AT. I also believe it is a good idea in general, because—unless some obscure research has escaped my scrutiny—there have been no previous attempts to bring the concept of AT into the realm of academia *as such*. I get a quasi-direct confirmation to my assumption by reading the very interesting meta-analysis provided in Bauer and Brown (2014: 347–349), which, on a sample of 53 papers about AT's, considers the recurrence of "indicators of appropriateness", that is, items/concepts/strategies employed in the assessment of an AT. There are altogether 47 indicators found, ranging from very recurrent ("Community input" pops up in 51 papers out of 53, "Affordability" in 36, etc.) to very occasional ones (e.g., "low energy", "reliability" and others, which are mentioned in only 2 papers each). Now, these data are not necessarily significant in relation to the amount of mentions they get in the sample's articles. What is more interesting is the (47) indicators that *are present*, and those that are not. And, among those that are not there, there is every single indicator (research, scholarship, academic input, etc.) that we may relate to an academic activity.

Therefore: how may this newly-proposed connection work?

When we think about both Schumacher's and Papanek's points, we can immediately connect with the "Keep it R.E.A.L." principle that I have suggested in point 3 of the manifesto. Conclusions are very similar, if not identical: the humanistic scholarship that is unattached to reality, abstract, indirect, pretentious, dispersive and unfeasible is clearly an inappropriate technology for academic research. Its unsustainability, uncontrolled "size" and independence from the (social, ethical, cultural, etc.) context, is exactly at the basis of the crisis, which is nothing other than an increased, pathological distance from the "real world" and the "human scale".

Most interesting of all, as far as our purposes are concerned, is Wicklein 1998 and 2004, which contain a 7-point classification of the so-called "Criteria to judge the appropriateness of technology", which I again would like to connect (sometimes literally, sometimes metaphorically—but not less significantly) with our program of reform for humanities. These are (all quotes are from Wicklein 2004):

1. **Systems-Independence**, related to "the ability of the technological device to stand alone, to do the job with few or no other supporting facilities or devices to aid in its function". In other words, we cannot evaluate a technology by the final product only, but we need to take into account all the supporting materials and equipment necessary for its full operation. Those have to be already in place.
   When it comes to humanities, the ability to exist as a stand-alone technology is a specific requirement of the more general concept of sustainability. A form of research that is fully operative "in place" is one that is not dispersive at both the methodological and theoretical level. If, for instance, a philosophical investigation on the origins of life reaches a point in which science is denied its function as "support material", and, say, religion takes that function (through creationism, intelligent design, or else), then we witness a similar situation as the product that cannot be operative "in place", but needs equipment from somewhere else. Just like those gadgets that are very fine as long as they work, but then as soon as a single, small piece starts malfunctioning, you need to send the whole thing to a technical support service on the other side of the planet.

2. **Image of Modernity**. This is less trivial than it may read at first glimpse. An "image of modernity" is not a marketing strategy of making something sound "new" (as Numanities themselves). "There is an innate desire within most of humankind to feel important and be perceived as worthwhile. It follows, therefore, that an image of being modern is important to the success of any technology. People must believe that a technological device brings with it a degree of sophistication which can elevate the user's social status as well as meet a basic human need". And: "All people have dignity and must believe that their dignity will remain intact even if they are poor or living in an underdeveloped area. Therefore, the designers of appropriate technology must incorporate an image of modernity into whatever device they develop if it is going to meet with acceptance by the people who can benefit from it the most".
   This is an important concept that gives the question of "novelty" and "modernity" a whole new twist that now relates directly with dignity and values. If we want "new" humanities is also because, as we have repeated several times already, we want to restore dignity to them.
3. **Individual Technology versus Collective Technology**. This criterion is related to the societal/cultural context in which a given technology will be established. "Some cultures advocate a strong commitment to the group process where the good of the whole is held in higher esteem than individual accomplishments. Other cultures place a high priority on individual responsibility and accomplishment. These factors should be considered in detail when designing appropriate technology because they will contribute to the success or failure of any given device or strategy. If a given cultural group has a strong allegiance to the community or region then the technology may be more system dependent, where the overall group could take a greater responsibility for the operation of a larger system".
   The application of this criterion to the humanities is quite self-evident, and should serve as a reminder to both scholars and institutions. Particularly the latter, in recent times, have become perhaps too exclusively obsessed with "collective technologies" in academic activities, forgetting—in fact failing—to create conditions wherein single scholars/teachers can make "a difference" by their individual skills and ideas. In point 2 of the manifesto, I mentioned this problem in relation to institutional restrictions in teaching programs.
4. **Cost of Technology**. That of course means "affordability", and we cannot but make another plea for forms of humanistic scholarship that are understandable in their concreteness and real world/human scale engagement. However, provocative this may sound, metaphysical, hardcore-postmodernist research has, in effect, a scientifically "costly" and unworthy quality-price balance.
5. **Risk Factor**. This refers to the degree of success and failure that the development of any technology (and produced result) carries with it. "Because of the nature of the device and the location in which it will be implemented the risk factor for the success of appropriate technologies […] is of major concern and should be considered an important criteria for development". There are two

types of risk that one must consider: internal and external. The former relates to how a technology will fit the local production system; the latter concerns whatever supportive system(s) may be required to have the technology properly operating. Of course, "although the risk factor must be taken into consideration when developing appropriate technology it would be economically and politically unhealthy to try to remove all risks. Some risks are healthy to the growth and development of locally implemented appropriate technology".

A totally safe form of humanistic scholarship is not worthwhile being pursued. Keeping in mind the terminology of the Boston Consulting Group, it can be safely said that humanities have offered their best to the history of knowledge when they tackled some delicate "question marks" or insisted in promoting ethically-significant "dogs" (as opposed to the safer "cash cows" or "stars"). The aim is certainly to make humanities system-independent (as we have seen in point 1 of this list), but that does not mean a full removal of the risks. At the same time, an eye should be kept also on the "cost" of the enterprise, so the element of rationality is also called into question. "Rational risks", to my mind, is a good (appropriate) way of calling this process.

6. **Evolutionary Capacity of Technology**. This consists of "the ability of the appropriate technology to continue to develop and expand beyond its originally intended function. […] If the appropriate device is static (i.e., performs one function and cannot be altered) then although it may provide for a basic need at the present time it will be a relatively short lived solution to a much larger problem. Wherever and whenever possible it is preferred that the appropriate technology allow for (i.e., have design characteristics) a continuation of development. That is, to have the capability to expand and be reconfigured to accomplish a higher volume of work and/or more sophisticated production processes".

This is a very important point that, like the third one on this list, calls for a plea not only to the community of scholars, but also to their affiliated institutions. And it calls again for a reflection on "time" (as already discussed in the second entry of the manifesto), and for the BCG Matrix jargon. More and more institutions, indeed, seem only to be content with investments of the "cash cows" type (and when I say investment, I do not only mean financial ones). Plenty of actions (say, a degree program, a research project, a development strategy…) are launched with enormous time pressure that allows only for one option: immediate success. When that is not achieved, the action is erased, and replaced with another one—with a significant loss of time, energy and—again—financial resources. The new action has indeed its own costs, learning curves, management problems, and so forth: it does not just *continue* from where the previous action had failed.

To take a significant example, in the universities I work or worked for, I witnessed more than one degree program killed after one year or two, because of a lack of students. Those programs had required important investments in terms of working hours, human resources and money, but they were not given the chance of a slow (or even false) start. The order was to close them, and one is always left wondering if the costs of starting a new program from scratch are really inferior to the losses of that old program's unsuccessful first year. In particular, that attitude shows a narrow-minded analysis of a problem which clearly has many variables at play: the program fails, therefore it is a bad program. What about bad timing, bad advertising, bad development, bad finalization…? And what if the program is simply not a cash cow, but an investment that will grow slowly and is doomed to become very solid?

There is a reason why a classic time frame for an activity's development is (or used to be) five years. I am aware I have introduced too many puns and acronyms in this book already (and still have more to display), but if another one is allowed, I would exactly use the word F.R.A.M.E.:

- F (the first year) is for **Fail**. New ideas need to have the chance to fail at the beginning. It is a basic right for anything new. If my young son learns to ride a bicycle without training wheels, he has an absolute right to fall, and nothing is more useful than falling, at this stage.
- R (the second year) is for **Repair**. With the right for failure comes also the duty to repair. At this stage, the action is under analysis and assessment: it has to prove that it has understood the initial mistakes, and can react to and fix them. It is a crucial moment in development. My son, now, puts a particular effort in understanding what, in his pedaling, was wrong and how he can improve. The academic institution has to be present in the action, and evaluate if the repairing stage is proceeding seriously and efficiently. If not, there are the conditions to erase the action.
- A (the third year) is for **Achieve**. If the "repair" stage was conducted properly, the third year will be the year when one can reap the fruits. Now, my son can stand on the bicycle and pedal correctly. If, on the contrary, there is no "achievement", the institution has again the right to erase the action. As we can see, even in a 5-year plan development, there is no need to wait 5 years if the conditions are not ideal. The important point is not to invalidate an action after a first initial failure. Depending on the development of events, an institution can intervene in the second or the third year. Not later, and most of all, not earlier.
- M (the fourth year) is for **Maintain**. To achieve success is already difficult, but to maintain it is even more complicated. The fourth year is a maturity test: keeping up with the example of the degree program, the success in the students' enrollment is as bound to circumstantial variables as the failure. If the program had failed because, say, it was "a bad year" (whatever that means: financial crisis, temporary unfashionableness of a given topic, etc.), it may also have succeeded for the same, but opposite, reasons (a "good year" of financial boom, fashionableness of a given topic, etc.). Back to the boy learning how to cycle, this is the time for him to show that he can now ride continuously, for a longer period, and so forth.
- E (the fifth year) is for **Expand**. Now, finally, we can apply the "Evolutionary Capacity of the Technology". Now it is the time for the action "to develop and expand beyond its originally intended function". Now the boy can ride bigger

bicycles, on different grounds, different weather conditions, he can even participate in some race or competition…

7. **Single-Purpose and Multi-Purpose Technology**. Not only should technologies be able to expand (see point 6). The possibility that they can be used in more than one application must be also considered. A technological device may be conceived with multiple purposes. The key-word, here, is eclecticism, and certainly one is reminded of the need to establish credible levels of inter- multi- cross- and transdisciplinarity for the humanities. Also, the idea that research can serve as "open source" for more research is an important aspect (Fig. 2.2) for a summary of the main characteristics of appropriate technologies.

I make no mystery of the fact that I consider this point of the manifesto a most crucial one, perhaps the most important of the whole project. On the occasion of the first International Congress of Numanities (Kaunas 2–7 June 2014), I delivered a kind of introductory lecture to let my colleagues know the main programmatic features of this new concept we were proposing. By no coincidence, I took a decision that the main focus had to be appropriate technologies. A few months later, I was invited to Seattle by the Semiotic Society of America, I insisted on the same points.

Of course, Numanities as such have to go a long, long way before assuming a defined and recognizable identity: in one of my case studies, later in this book, I quoted one of my favorite Italian songwriters, Sergio Caputo, who said that "songs

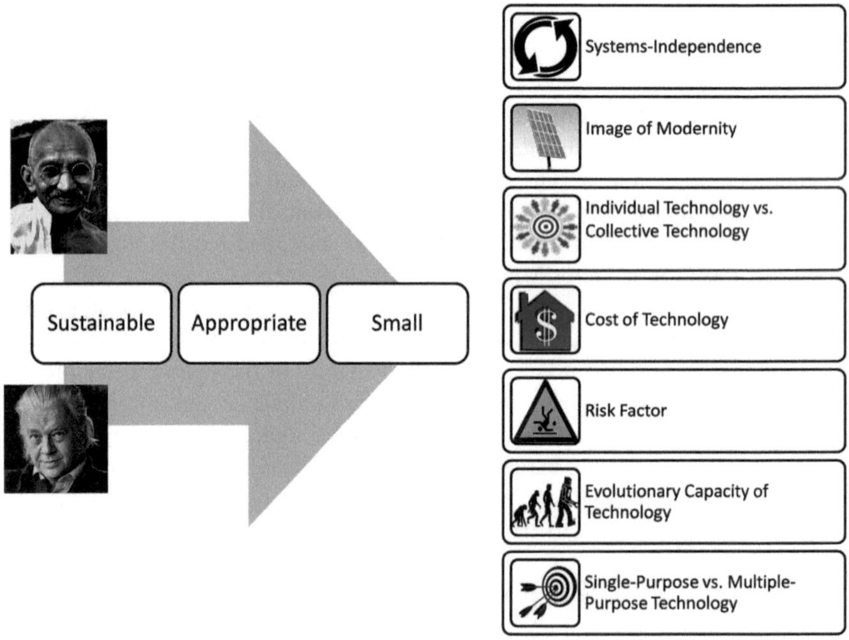

Fig. 2.2 Characteristics and criteria of an appropriate technology

are like boxers". That is: they need to be hit several times, and in public, before they take their final shape. I think the same applies to a new academic discipline (or platform/forum, as Numanities intend to be). Even so, however, while I am ready to cope with the fact that other points of this manifesto may be subject to redefinition, second thoughts, or at least may not be applicable all the time (e.g., the idea of re-establishing a dialogue with institutions may end up in a failure; one may not need to engage in full interdisciplinarity every single time, etc.), I do not seem to be able to make any sense out of this whole enterprise if we do not take a firm and durable stand on the fact that humanities need to pursue and produce quality and dignity of life. Without that, they are closer to self-indulgence than to science—and too bad if someone gets irritated by this statement.

Anyway, given also its tones, non-academic passages and rather peculiar example of appropriate technology, I decided to write here the full transcript of the key-note I gave in Seattle (which, unlike the one given on Kaunas, had less of a stress on the more institutional aspects—which are anyway disseminated throughout the present monograph—and perhaps a bit more light tones):

Dear friends, dear colleagues, ladies and gentlemen

About 10 months ago, when we opened the International Semiotics Institute in Kaunas University of Technology, we launched a "research concept" called Numanities—as in New Humanities. The goal is to establish an interdisciplinary platform that would mostly investigate the position of humanities in nowadays society, its potentials, its challenges and its relation with technologies. It is possible that some of you have read either some of the short articles I published here and there on the subject. And it is also possible that you may be already aware of some of the comments received by this new paradigm, so far. Humanists often have a reputation for being slightly technophobic, and indeed a recurrent type of remark on these articles was that Numanities were a sort of uncritical celebration of technologies and a wish for humanities to become totally devoted to them, much to the detriment of what was perceived as their "real" identity.

Now, it is not my wish to be polemical with these interpretations, but I honestly have the feeling that they must have come from colleagues who either did not really read those documents, or they did not really understand them.

In the paradigm of Numanities, in fact, this question is dealt with very carefully, and it was made loud and clear that the way we need to envision the relationship between humanities and technologies is for the former to "monitor" and "assess" the latter, and make sure that they are always compatible with quality and dignity of life.

However, there is a serious chance that I was not particularly clear in "defining" technologies, and in particular this relationship, and that—at the end of the day—corresponds to the final goal of Numanities: that is, to discuss the actual position of humanities within contemporary society.

So I decided to clarify these aims on this occasion, and for this reason I came here to talk about… my middle age crisis.

Seriously!

I turned 40 just a couple of months ago, but I had signs of this crisis already last summer. Except that it had a rather unusual manifestation on me. I had heard that what normally happens is that one wishes to "be younger", so they start to do, say and most of all wear *things that are more appropriate for someone half their age. On the contrary, what happened to me was a vital urge to be and look "adult". I'll spare you the details of this process, which involves several different steps and actions, but you can certainly imagine that, on occasions, I must have looked a bit pathetic, particularly to my wife, who started looking at me with a mix of amusement and compassion.*

Then again, in that case I have at least two theories. One theory is that we are all a bit unprepared for to changes, particularly those that occur to people we know well. We have a certain image of them, and we were always comfortable with it: how dare they mess up our equilibrium? I guess you know what I mean: for instance, when we meet friends or relatives that we haven't met for years, they always expect to see that "version of ourselves" that used to exist back then, *but obviously doesn't exist anymore. So we get to hear things like "I don't recognize you anymore", "what happened to you?" and the likes, as if "to be yourself" means to get stuck in one specific space-time point.*

The other theory, a more chauvinistic one, is that my wife, like all women, take only Brad Pitt seriously. Or a few other men of that sort. When the first wrinkles appear, normal men like me get old. *Brad Pitt instead, becomes even more sexy. When the normal man dresses more elegantly,* he's not himself anymore. *Brad Pitt becomes even more sexy. And so on.*

Anyway, let's go back to business.

Fortunately, during this process of "becoming adult", a few interesting things happened as well. For instance, I became very attracted by traditional shaving, the one you do with a steel safety razor, shaving soap and a brush. Then, since I'm crazy, I also bought myself the alum block to fix little haemorrhages and even some of those aftershaves that old-time barbershops used to have: you can still find them on the internet and they have an exquisite perfume.

In sum: a radical, fanatic full immersion. *Perhaps my wife is right. Perhaps even Brad Pitt would look ridiculous in similar conditions. Or then again, maybe he will look even more sexy, with this "real man" accessory.*

Let me explain.

All my life, since when I started shaving, I had irritation problems to my skin, both with disposable and electric razors (although the latter were admittedly better, but then—on the other hand—the quality of their shaving was worse). The legend, massively fuelled by advertising, was (and is) that traditional razors are much worse: not only do they irritate our skin more, but they also regularly cut it and make it bleed. The blade goes straight on the skin, they do not have an ergonomic shape, they use an "old technology" (see? We are getting to the point!), and this technology was overcome by more recent research… for more than 20 years, all this pressure kept me very far from classic shaving: I was convinced that, if I tried that, my face would have soon looked like I was shaved by Sweeney Todd.

Therefore: I first tried the disposable razors and, after a few years, I turned to the electric one.

Then comes the middle age crisis: I end up reading some stuff about classic shaving (there are even blogs and forums on the net), and one of the recurrent comments is that the alleged "danger" of this practice is a total myth. On the contrary: these forums claim that it is all these special blades, super-blades, triple-blades, filters, soothing material, and all the fancy stuff in the new razors to make things worse. First because of the variety itself, which is actually distressful for the skin, rather than helpful, and second because the main irritating factors are the actual leftovers of the shave (small fragments of hair and cream) that remain stuck on the blades and are difficult to remove: this is not a problem in classic blades, which can be easily and totally cleaned, while it is an issue in disposable razors (which is why we cannot use them many times).

I find this information rather convincing, but still there is the other problem*: haemorrhages. The terror of traditional razors is celebrated by dozens of books and movies: how to forget that sequence in the movie* The Wall *when the character Pink shaves himself to total baldness and in the process he bleeds like rain?*

However, even in that case, the material I read offers reasonable explanations: the point, apparently, is simply the know-how. One has to learn how to shave: you need a light but firm handling of the razor, you have to shave with short movements, you have to incline the razor to about 45 degrees from the skin, you have to learn the beard "map" of your face, which is different from person to person (the beard doesn't grow in one direction only), and so forth. My goodness! Too much work! This kind of things is normally enough for me to lose motivation: at this point, I am very close to sticking to my good (old?) electric razor.

But what about my middle age crisis? Come on, I have to give it a try. So: I buy all these classic products for classic shaving, and I start.

And of course it's a total disaster! If I had shaved with a cactus I would be bleeding less. Thankfully I bought also the alum block: that one, at least, works very well (by the way, how come they nearly don't sell it anymore?).

Also the irritation problem is not really solved: my neck has the traditional purple colour that it has always had every morning for the last 20 years.

Last but not least, it takes me four times as long to finish, as compared to electric shaving.

However, I don't give up. Time passes, I start "taking the measures" of my face, I learn the right movements, I am less intimidated, and the use of the alum block is reduced every day, up to disappearing completely. I am not bleeding anymore!

Irritation, too, almost disappears, and is far less than any of the methods I used before. Plus, my shaving is perfect: every day my skin feels like my son's bottom… totally smooth and soft! Beforehand, no matter how hard I would try, there would always be a couple of spots that I could use to light a match, if you know what I mean.

Finally, thanks to my fanaticism, I also smell great, in a very classy, classic and cosy way.

The only trouble left is that I haven't improved with time: I still need some 20 min to complete my shaving, as opposed to the 5 circa of my "electric" past.

By now, after almost a year, I totally master the practice, and it is a very pleasant activity that I perform in total accordance with my middle age crisis: it feels nice, adult and, yes, it feels "manly". I still don't feel like Brad Pitt, but then again mother nature didn't help there. In fact, there is a lot of talking, these days, about the crisis of masculinity and manliness. It has been something going on for a few years now. You must have noticed, for instance, how many men's magazines are being published nowadays, GQ, Men's health, Men's fitness, etc. *And they are all of the same, repeating, boring type: how to work out, how to do business, how to win a woman. That, my friends, is* really *pathetic.*

Now. Shall we get to humanities and technologies, finally? Of course we were always talking about them, and I am sure that some of you have already understood my point.

While approaching classic shaving, indeed, I understood perfectly the reason why modern society has demonized it, through the production and the promotion of other, supposedly advanced, *technologies. The fact is: classic shaving is less irritating, less harmful and much more effective than more recent practices.*

However:

1. *It has a* learning curve. *You've got to learn to shave properly. The improvement of your results will not occur by switching from one technology to a more advanced one. And it will not occur by buying more products, that soothe here, facilitate there, nourish this, enrich that. The improvement will occur with the* same *object, and by investing time and knowledge,* not *money.*
2. *Classic shaving is ecologically sustainable. It doesn't consume electric energy and it has only one small part to change every now and then. That part, the blade, is made with metal only, and it comes with a small paper packaging. There is no plastic involved and no chemical substances of any sort. In fact, you can even buy a tool for sharpening the blades, so that they can last much longer. That, too, is not possible with disposable razors.*
3. *Classic shaving is also economically sustainable. The expenses for purchase and maintenance are far inferior to more recent methods. I bought an excellent safety razor (suggested by most forums), for only 40 euros: it is a Merkur 34C and it is German—which alone is a good thing, when it comes to technologies! A package of five excellent blades (which last more than a month), costs less than a euro. For an excellent electric razor, I need well over 100 euros; the energy I consume for shaving is more than one euro per month, and let's not even start with changing pieces, when you need them. As you know, they cost so much, that they make you want to buy a new razor altogether. Disposable razors, on the other hand, are even more expensive, particularly those new fancy ones that come straight from Sci-fi movies. Apart from this, the economic advantage of safety razors is also in their eternity. Unless I lose it somewhere, my Merkur will last forever, and I can also hand it to my son, providing it with an additional affective value that other products will never have.*

4. *Finally, classic shaving requires time. To shave becomes a specific occupation, with its own identity: a significant amount of our time (20 min in my case) has to be devoted to—to what? Well, to ourselves, to our own care. It may be nice to return to the 5 min of electric shaving, yet even this "waste of time" has become for me meaningful, and I shall soon explain why... although you may have guessed it already.*

So: knowledge, ecology, economy, time. Four undisputable pillars in the life of a person and his/her community. Four pillars that modern society (particularly in its most extremely capitalistic incarnation) wants to manage on our behalf, instead of us. Knowledge is more and more produced in a superficial way, particularly the knowledge of daily life. It is important that we don't know too much, so that it becomes possible to "create new needs": one way is to increment the speed of life, not giving us time to understand whether a given object is or not suitable, appropriate for us. There is no time to understand that, because in the meanwhile there is already an "update" or an "upgrade" available on the market.

Plus, knowledge is the noblest form of "property". We do not really own *anything until we* know *it well. To take just one example, as long as we buy a smartphone with 100 different functions, of which we understand and use only 10, we don't really* own *it: it still belongs to the company that sold it to us, and that company will soon sell us the newest version, with 120 functions, of which we will use only 11, losing even more property of the item.*

Then: sobriety and sustainability, both ecological and economical, are enemies of profit and commodification (at least of "quick and easy profit"): things must be soon broken, changed, improved, in order to generate profit. A sober/sustainable item is good for me and for the environment, not for those who want to sell me a product I don't need.

Finally, time. If there is one thing that is still worth more than money, that is time. The moment we can afford *20 min instead of 5 for taking care of ourselves, is literally a moment of richness. Someone, somewhere, wants to take that time away from us, so that we can use it for* their *purposes. If we can spend 5 min to shave and then we use the remaining 15 for a shower, or for a book or to play with our children, then we still* own *that time. But if we need those extra 15 min to go to a shop (perhaps to buy a new disposable razor), then that time is lost forever, it doesn't belong to us anymore. To retake possession of time, as much time as we can, means literally to* live longer. *In football there is a distinction between the "length of the game" (which is notoriously 90 min) and the "effective time", which is calculated by subtracting all the breaks in the game from the overall length: fouls, substitutions, celebration after goals,* etc. *It is normally believed that you had a lively game when the effective timing was at least 60 min: it means that a lot of playing was made. Under that threshold, it means that too much time was wasted. There we go: we always calculate our life by length,* how long we lived*: 80 years? 90 years? But we do not consider the "effective timing" of our life, which, on the contrary, would really tell us how* much *we lived.*

Most of all, however, these four pillars correspond, with great precision, to the concept of "appropriate technology", which, in its most accurate definition, was formulated in the 1970s by the economist Ernst Schumacher, in his groundbreaking book Small Is Beautiful, *but which—as a principle—was mostly one of Gandhi's ideas for the economic and psychological independence of the Indian people. Nowadays, "sustainable development" has become a more fashionable expression to describe more or less the same basic ideas, but I prefer to keep the original formulation, so that it remains clear that I am still talking* about *technologies, and not* against *them. In Schumacher's formulation, an appropriate technology has to be environmentally sustainable, socially appropriate, and must be animated primarily by three values: health, beauty and permanence. This is what those four pillars are about: an appropriate technology responds to* my *needs and* my *wishes, not the ones artificially created by someone else. It is a bit funny how so many communities, Lithuania included, see "progress" as a dichotomy between innovation versus conservation. We have the innovators and the conservatives. But progress is not about going forwards or backwards. Progress is about going* towards.

Now you can see where I place humanities in society, and where I would envision a paradigm like Numanities. The humanities have always had this role of thermometer of our quality and dignity of life: as a matter of fact, they were chiefly responsible for creating *quality and dignity. Arts, culture, philosophy, ethics… these are the fields that go* towards *us, as individuals and as community. Humanities have a place for both* forwards *and* backwards, *as long as they go* towards.

Humanities have a specific obligation not to be *technophobic: on the contrary they have to constantly interact with technologies, because they are the only approach to knowledge that can actually* critically monitor *technologies, and make sure that the* appropriate *ones are welcome, while the inappropriate ones get smashed and thrown back to the face of the lousy business people who want to decide in our place.*

It is a question of knowledge, ecology, economy and time.

Thank you for your attention.

To conclude, I have employed several catchy expressions in this chapter: I spoke about progress that should go "towards", rather than backwards or forwards, I spoke about knowledge, ecology, economy and time, and so forth. However, as catchy as they all are (and I like them to be), I really do not mean these expressions in a rhetorical, slogan-like, sense. That progress can and should not be assessed as a straight (diachronic) line is something that I have tried to prove in various ways. But let me still mention another example, my favorite one (also because it has a happy ending): public washrooms.

Roughly speaking, since the concern for hygiene made it clear that simple fabric towels could not be used, restrooms in public places have offered their customers four different ways to dry their hands: paper towels, fabric towels in a roller, warm air dryers and, much more recently (the first one having appeared in UK in 2006), the so-called airblades (those with an air jet speed of over 600 km/h that look like they could disintegrate your hands at any moment). Achieved a decent (although not equal) hygiene level for all of them, the main concern around these technologies became of course their environmental impact. Old and new technologies have been used, here, alone or combined: a paper napkin, for example, can be taken from a simple box or from an infrared dispenser. Now: does anyone know which of these technologies has the lowest environmental impact, all considered?[8] Was it better to use more recent strategies or rely on the good old ones? Is it better to waste paper or energy for warm air? Is it worse to use chemical products to wash fabric towels that will be reused, or to dispose paper, albeit recycled one? But, most of all, was "progress", in the development of these technologies, going in a straight line (be it in the forwards or backwards direction)? Was there any "towards" progress at all?

The answer, documented in an MIT study (fully available online at http://msl.mit.edu/publications/HandDryingLCA-Report.pdf), shows exactly how progress is a bumpy, curvy and twisting road. Summing all the six parameters of the LCA (see footnote 8), the worst thing we can do to the planet, when drying our hands in a public restroom, is to use paper towels (36 impact points altogether). Warm air dryers are just a trifle better (35), but certainly in no significant manner (no real progress shown). Things timidly improve if we use recycled paper towels—as opposed to virgin ones (29), or—even better—if we use a fabric towel roller (24). Significant progress in the warm air drying technology occurred in the 1990s when the Japanese Xlerator shortened the drying time from the customary (and unbearable, particularly in summer) one-or-more minute of the standard dryers into ca. 15 seconds (the total impact now was brought down to 19). But it was only when Dyson airblades appeared (particularly plastic ones, as opposed to aluminium ones, which score higher in water consumption) that a minimal environmental impact was reached. A plastic airblade scores only 1 in each of the six LCA parameters, totalizing a very low 6. In other words, progress had to renounce the backwards or forwards modes, in order to find its way "towards" us. It is nice that this particular example displays a success story: the most recent technology ended up being the best one, in both environmental impact and quality of life, since airblades dry our hands in only 10 seconds. But of course we are exposed daily to thousands of technologies that—in their development—worsened our quality and dignity of life, while deceiving us with promises of cheaper prices or higher luxury. Millions of technological stories, at the present state of things, do not stand out for their brilliance: some brought more problems than solutions (e.g., SUV cars), some ended up being used the wrong (and/or evil) way (e.g., Zyklon B), some were just very, very stupid ideas (like the short-lived motor powered roller skates, which

[8]The scientific method of analysing the environmental impact of a product is called Life Cycle Assessment (LCA) and includes six parameters of evaluation: the global warming potential, the human health, the ecosystem quality, the cumulative energy demand, the water consumption and the land occupation. A scale from 1 to 7 (from the lowest to the highest impact) is assigned to each parameter. This way, different products or technologies which have the same purpose can be directly compared and assessed, taking into account every single step of their production, and related impact.

caused hundreds of accidents), and finally some are so shamefully uncivilized that one should not even address any other question in relation to their use and simply remove them asap (e.g., intensive farming).

As we said, progress is a bumpy, curvy and twisting road, and will always be so. What we need to do is making it "our" road—a road where twists, curves and bumps are not obstacles, but charming features that improve the landscape.

2.6 Promote Values That Other Fields of Knowledge Are Not Capable of Formulating

This point of the manifesto is, by all means, not intended as a manifestation of superiority, but—on the contrary—it is a claim for nurturing identities in the most appropriate way. If natural sciences study and create what makes life *possible*, humanities study and create what makes life *worthwhile*. Creativity, beauty, quality of life, dignity of life, empathy, tolerance, culture, reflection… this is what humanities do best, and what humanities are mostly for. In a very appropriate pick of words, the *Heart of the Matter* report affirms that humanities are "the keeper of the republic" (Broadhead et al. 2013: 9).

If society and institutions have become reluctant to understand the importance of these values, then humanities need to find new ways (or 'well-temper' the old ones) to convey the message. It cannot be only society's fault: yes, events went in such a direction that humanities were deprived of many of their legitimate rights, but it is also true that humanities themselves became too self-confident and self-indulgent towards their traditional centrality in society. Such an attitude, besides being wrong in general, was particularly wrong in relation to the critical and self-critical nature of humanist thinking, which should have been the *first* to reject any inclination towards dogmatism.

Maturity is shown in moments of difficulty: these are hard times for the humanities, thus this is the right moment to show strength, resilience, adulthood and adaptation skills, keeping in mind that in evolutionism "adaptation" is not surrendering, but rather the highest form of physical and mental development of an organism.

Moreover, this is hardly the first time that humanities, or any other field, are undergoing a crisis of any sort. Natural sciences, because of their wonderful work-in-progress constitution, have faced dozens of epochal changes, where a radical rethinking of their paradigm and actions was needed: it cannot have been easy to believe for centuries that Earth was a still body at the very center of the universe and then suddenly face the Galilean revolution; or to believe in the uniqueness of human beings and then suddenly face the Darwinian revolution. However, the humanities, too, can take a look at their own past and find comfort in how effectively they managed to overcome their own crises: the transition from the Middle Ages to the Renaissance, from patronage to entrepreneurship in arts, etc., not forgetting that the above-mentioned crises in natural sciences had a huge impact on humanities too.

In the introductive part of this book, we mentioned that the crisis of humanities involves also their identity as producers and *promoters* of knowledge. Indeed, part of the enormous heritage that humanities have delivered to the world consists also in constantly stressing the importance of knowledge and culture. They did not just provide material to acquire them: they were also the main PR of them. That is nowadays more important than ever, as the need for knowledge and culture has become increasingly marginal in modern society (and that, among other things, is at the basis of a large number of problems in the world, starting from the various manifestations and typologies of prejudice and intolerance—all direct offspring of ignorance).

Somewhere along the way, humanities lost part of their ability to convey a convincing image of "knowledge and culture". Students are growing uninterested, even hostile, to them: they find them impractical, not responsive to their needs: once again, it may be society's fault, but to complain and not move a finger about it is not going to help. Whether the solution is to change attitudes or to fight concretely and restore the old status quo, that role *must be* recovered.

We have already mentioned (in point 3 of this manifesto) Dilthey's definition of the humanities as an "understanding" process (as complementary to natural sciences' inclination to "explaining"): that is already a founding element of uniqueness for the humanities. However, in order to deepen the issue, Nussbaum's treatise (2010: 33–34) is an excellent framework to understand what specifically makes (or could make) the humanities an added value for knowledge production. Nussbaum emphasizes seven specific abilities:

1. The ability to think well about political issues affecting the nation, to examine, reflect, argue, and debate, deferring to neither tradition nor authority.
2. The ability to recognize fellow citizens as people with equal rights, even though they may be different in race, religion, gender or sexuality: to look at them with respect, as ends, not just as tools to be manipulated for one's own profit.
3. The ability to have concern for the lives of others, to grasp what policies of many types mean for the opportunities and experiences of one's fellow citizens, of many types, and for people outside one's own nation.
4. The ability to imagine well a variety of complex issues affecting the story of a human life as it unfolds: to think about childhood, adolescence, family relationships, illness, death and much more in a way informed by an understanding of a wide range of human stories, not just by aggregate data.
5. The ability to judge political leaders critically, but with an informed and realistic sense of the possibilities available to them.
6. The ability to think about the good of the nation as a whole, not just that of one's own local group.

7. The ability to see one's own nation, in turn, as a part of a complicated world order in which issues of many kinds require intelligent transnational deliberation for their resolution.

In principle, one could simply subscribe to these seven points without even bothering to comment on them: they *are,* indeed, an exhaustive description of what the humanities can do (*better than*, *more than*, or sometimes *instead of*) other fields of knowledge. At the same time, as we have seen in point 1 of the manifesto, it is one of Numanities' aims to act as a wider discussion forum for different approaches and strategies to reform and reevaluate the humanities. Perhaps, thus, in the light of these different paradigms, some remarks to Nussbaum's list are possible:

(1a) *The ability to think well about political issues affecting the nation, to examine, reflect, argue, and debate, deferring to neither tradition nor authority.* There is hardly anything to say about this point, except that Numanities are *particularly* interested in this cultural challenge to aprioristic "tradition" and "authority". In the previous points of the manifesto, I had the opportunity to emphasize (and possibly dismiss) the apparent paradox of Numanities: defending integrity, excellence and solid preparation in the humanities on the one hand, and criticizing technophobia, "pastness" (we shall see this word defined by Fredric Jameson in point 7 of the manifesto) and tradition for tradition's sake attitudes on the other. Please, let me repeat: there is no paradox here. The humanities should claim their authority and authoritativeness on the basis—as correctly pointed out by Nussbaum—of their ability to "examine", "reflect", "argue" and "debate". Quite simply, that can equally praise or challenge tradition, technology, the past, the future, the present… it merely depends on the cases. The humanistic catalogue of skills is *exactly* meant to discern and discuss critically.

(2a) *The ability to recognize fellow citizens as people with equal rights, even though they may be different in race, religion, gender or sexuality: to look at them with respect, as ends, not just as tools to be manipulated for one's own profit.* We again can generally agree, here, but a remark must be made on how natural sciences can bring important added values to this vision. For instance, the humanities—in most of their incarnation, including Nussbaum's[9]—tend far too often to exclude the "non-human" from their global ethical project of equality, respect and personhood, on the basis of outdated notions/perceptions that are not anymore in step with what science reveals as some states, governments and institutions are beginning to acknowledge. It is—I repeat—one of the great assets brought by the post-humanistic approach that we have highlighted in point 1 of the manifesto. If it is true that the humanities have the right predisposition to develop

[9]Although, it must be said, she displays remarkable competence on De Waal's studies on sympathy and compassion, definitely placing herself on a different level than the usual anthropocentric-speciesist cosmology developed by most humanists.

empathy, tolerance and understanding, it is also true that they applied (and assigned) these qualities only to and within humankind, not infrequently explaining this standpoint with emphatic statements of human superiority ("unlike animals, we are endowed with passion/culture/sensibility…"), that are at the same time a contradiction to those very qualities they should develop, and are a display of ethological/zoological ignorance.

(3a) *The ability to have concern for the lives of others, to grasp what policies of many types mean for the opportunities and experiences of one's fellow citizens, of many types, and for people outside one's own nation.* In this early stage of Numanities' development, we should again generally agree, provided the point made above, and provided the sensation that it was mostly the humanists themselves who chose a romantic-nationalist-individualist path (as I have emphasized already, and shall do so again in the next point of this manifesto), perhaps deviating from the discourse appeared and progressively developed in the period between 15th and 18th centuries. Once again, one may detect an apparent paradox, here, in a treatise, like the present one, which shows cultural sympathies for *both* the Renaissance's idea of humanism, and the post-humanistic approach that so much criticizes it. I can only reiterate on Nussbaum's first emphasized quality of the humanities: if the humanities can examine, reflect, argue and debate, they can also avoid generalizing and treating topics monolithically. The humanism of the Renaissance is not so much an emancipation of human beings from nature, as it is from God and superstition.

(4a) *The ability to imagine well a variety of complex issues affecting the story of a human life as it unfolds: to think about childhood, adolescence, family relationships, illness, death and much more in a way informed by an understanding of a wide range of human stories, not just by aggregate data.* Once again, while we generally agree on this, we cannot help noticing that "data aggregation" has been a favorite occupation among plenty of humanists, who—contradicting the first of Nussbaum's agenda points—treated tradition as authority, and generated numerous forms of data aggregation for aggregation's sake that, on balance, were not enormously valuable for the understanding of "complex issues affecting the story of a human life". I am not suggesting the removal, from the humanities' universe, of obsessed philologists who can devote their entire life to dig the world's archives in search of every single handwritten private letter of some dramatically-marginal erudite of the 16th century: however, I envision a very difficult collocation of such works in the development of Numanities, or at least a very hard time in understanding how on earth these kind of enterprises will help us understand and improve our society, if not in the general assumption that it contributes to expanding our knowledge (an assumption that, in an attempt to raise the bar of expectations towards humanities, should not be considered enough anymore).

(5a) *The ability to judge political leaders critically, but with an informed and realistic sense of the possibilities available to them.* We generally agree, but we still suspect that "realism" has ceased to be the forte of the humanities, which instead should activate a serious process of self-assessment, in order to reacquire it. I remind the readers of the previous points of the manifesto, and to the next one too, for a more thorough discussion on these aspects. It is my absolute conviction that the loss of realism is one of the main causes of the humanities' crisis.

(6a) *The ability to think about the good of the nation as a whole, not just that of one's own local group*. We generally agree, but we are not entirely sure about the choice of the concept of "nation" as the unit of measurement for an enlarged community. In fact, one may even see the same danger emphasized in the third entry of this list of comments: the problem of the current romantic-nationalist-individualist path chosen by the humanities may have something to do with this identification of the nation with "the great picture". While we all agree on the importance to "act locally", the "think globally" part of Greenpeace's celebrated motto may need to be more extended than the territory of a single country, and moreover involve not just "human people" but the whole ecosystem. Most of the world's problems that carry a local impact have global causes and global explanations. The Numanities' nation is the planet (emphasis on "planet", as opposed to "world"—as I said, we do not want only people in it).

(7a) *The ability to see one's own nation, in turn, as a part of a complicated world order in which issues of many kinds require intelligent transnational deliberation for their resolution*. This final point is certainly more accommodating than the previous one, as it finally sees "the nation" as an intermediate point of social awareness, and not as its goal.

These, at least these, are the added values that a humanistic approach brings to the process of creating knowledge in a community. The idea of "humanities as added value" (we shall see it later) is the main epistemological pillar of Numanities. With a common and committed effort, we need to convey the message that the humanities still display concretely irreplaceable components to this process. In fact, it is not like such components have disappeared: they are merely less visible and—fatally—less appreciated institutionally. But, for instance, if the majority of a state leadership still consists of people with a humanistic background (as the collective "4Humanities" has underlined—see the introduction to this monograph), it can only mean that this "added value" is still operative and effective.

Finally, as a complement to Nussbaum's reflections, I shall take the liberty to suggest a possible program of "values" for the Numanities. How, in other words, we would intend to be an added value, in the current social and academic context. This formulation, in 8 points, has been a symbolic statement of purposes for the International Semiotics Institute since January 2014. As we interpret these eight values as something "we want to be proud of", we decided to elaborate them around the acronym (yet another one, I know!) of "PEACOCKS", the proud animals by definition.

The peacocks of Numanities are:

- P is for PASSION: humanists often distinguish themselves for a high degree of commitment and engagement. Doing things with love and enthusiasm is the first important step for excellence.
- E is for ETHICS: the humanities, as Nussbaum and others repeatedly remark and as I have specified also in this manifesto, seek to promote values of dignity and quality of life: tolerance, peace, equality, empathy, democracy, emancipation, etc.
- A is for APPLICATION: whatever we think or write, we should always try to apply it to the real world. If we cannot afford to reach a stage of concreteness from wherever we start our research work, then we should seriously question its validity.
- C is for CREATIVITY: traditionally that has tended to be the distinctive feature of the humanities. We need to keep and reinforce that: creativity means believing in and promoting imagination, art, ideas and vision.
- O is for OPENNESS: the humanities should develop more and more into an open-source and open-platform entity, devoted to diversity, inclusiveness and transparency. Openness is where the repeatedly-mentioned ability to "examine, reflect, argue and debate" finds its natural element.
- C is for CURIOSITY: a scholar is not a scholar if s/he is not driven by a genuine desire to look around, learn and understand. The humanities possess a natural inclination for exploration, but—in comparison to the natural sciences—they may have partly lost the value of open-minded, first-hand curiosity, increasingly replaced by a more speculative and mediated type.
- K is for KNOWLEDGE: that means information, communication, culture… Needless to say, this is what research is all about. The warning, here, is to never forget that, at the end of whatever scholarly enterprise, we should *produce* knowledge, not just question it.
- S is for SUSTAINABILITY: it stands for ecology, appropriate technologies, sobriety, non-invasiveness and "economy" in its truest sense. In the case of humanities, sustainability means two things: first—indeed—a plea to always pay attention to the impact of our research on society (making sure that such an impact is a positive and useful one); second, once again, it is a challenge to metaphysicalism and its tendency to cosmic (and therefore anti-economic) speculations (Fig. 2.3).

Playing with words, humanities will return to being an added value when they manage to add value to our society. We all secretly think that we do that already, and that the world would not survive without us, but the fact is, it is surviving alright. It is getting uglier, but it is surviving. Also—although I will make myself a few enemies here—we may want to admit that some of our work is not creating value at all, even though we like to think that it does. An important step to overcome the crisis is to emancipate ourselves from this self-indulgent conviction that nothing of what we do is questionable, while—more and more often—it actually is.

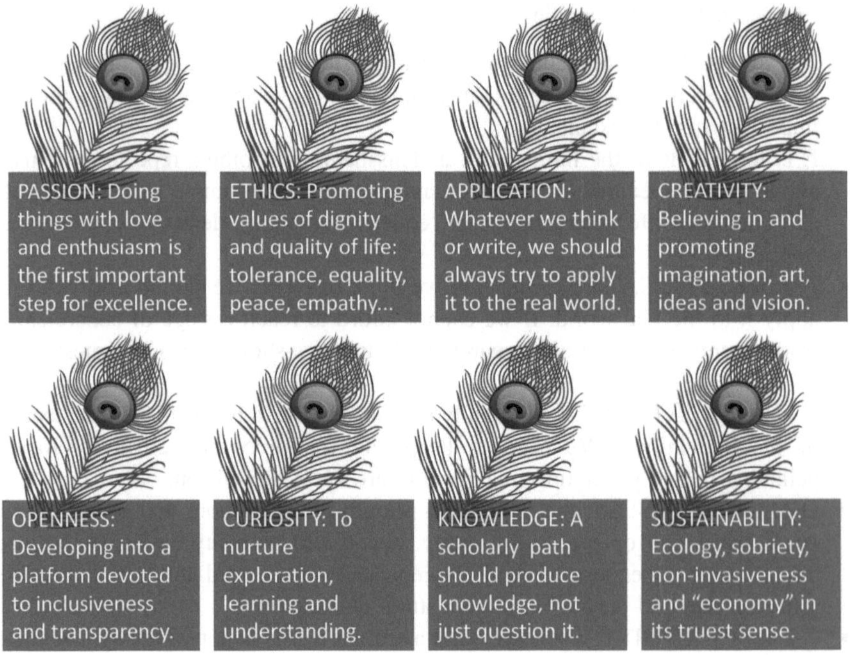

Fig. 2.3 Eight values (P.E.A.C.O.C.K.S.) pursued by Numanities

2.7 Strive for the Excellence of the Humanities

As we have seen before, Diana Sheets suggested that humanities have to restore "rigorous standards of academic excellence", and of course how can one disagree with this, generally speaking? In the current crisis of the humanities, these standards are more seriously threatened than ever. Changes in society aside, *this* is one of the prominent factors that are giving humanities a bad reputation: the *noise* of mediocre research has increased in volume, and therefore it is more and more difficult for institutions to spot excellence: the hyper-bureaucratic, time-consuming and annoying funding processes are certainly something to change radically and possibly abolish, in an ideal world (see point 2 of this manifesto).

However: have we ever thought that perhaps it is also *our* fault that things are so difficult and controlled? Have we considered that perhaps some (or many) of us are really not properly competent and prepared? What is our reply to *the system*, besides the refrain "Humanities are humanities, and therefore cannot be ignored" (which is a rather literal equivalent of "traditions are traditions")? How do we prove to the institutions that this or that research plan is really groundbreaking, or this or that heritage really needs to be preserved and studied in the particular manner we intend to?

Maybe, in the "good old days", humanists did not need to produce such evidence: the necessity of philosophical speculation, heritage preservation or else, was simply self-evident. Yes, but in the good old days there were not so many humanists struggling for one single piece of cake, simply because it was *more difficult* to be a humanist. Perhaps *all* good humanists had a prominent place in society, but only a few were good enough to access the club. Nowadays, the proportion between humanists and the available resources is totally unbalanced: humanists have increased to a ridiculous number, and resources, to make things worse, were in turn cut down. There is no doubt that they were reduced in unfair quantities, and that institutions have instead reinforced their support to openly profitable fields (without too much investigation of the scientific-cultural significance of each specific project), but it would be wrong to assume that this happened only because of the wild capitalistic mentality currently dominating in the world.

Modern societies have taken a sad, yet important decision: they have decided that culture, arts and philosophy are not endlessly sustainable: in fact, they are just sustainable in small amounts. Humanities were definitely caught by surprise by this decision: they understand that it is a wrong one, but they have not been able to produce valid counterarguments, besides those that tautologically run in circles, nor did they manage to adapt to the new picture (save some exceptions).

To take a firm stand against mediocrity, humanists need also to resist engaging in self-indulgent, narcissistic and—once more—unempirical and abusively-speculative forms of scholarship. One of my favorite examples in the category is post-modernism (or, rather, the degeneration of it that developed in academic circles between late 1980s and early 2000s). By now, post-modernism, at least part of it, is no longer untouchable, and that is certainly great news for anybody who is looking forward to a rebirth of humanities that would consistently reduce the whole package of skeptic-hypersubjective-relativistic-antiscientific *clichés* that have been practiced in the last few years. Lyotardian premises of the "depends-how-you-define-it" type, proved to be exceptionally powerful weapons against any attempt to develop a discussion in the forward mode. The request to define something that the post-modernist has already decided s/he will not accept as objective, demands a step backwards in the discussion; then, some time is spent on fighting over the possibility of actually defining that concept; then another tricky concept is found, and another step back is required. And so on, until the theorem of "infinite regress", which postulates that one can go back to point 0 from basically any point, is once again proved (point 0 being of course the impossibility to have anything remotely close to "truth" and "objective reality").

What is wrong (that is, far from "rigorous standards of excellence") in post-modernism can be summarized in at least the following points:

1. The paradigmatic post-modernist attack on (the possibility of achieving) value-free objectivity in scientific inquiry is probably the most discussed aspect, and has been wisely counter-argued by many authors. Their main message is well summarized by the anthropologist D'Andrade (1995), who states that such an impossibility is definitely not a good reason for an anthropologist (and, of

course, by extension, any other scientist) to renounce having objectivity as an ideal, and to get as close as possible to it. Objectivity, D'Andrade points out, is not *dehumanizing* nor is objectivity impossible: "Science works not because it produces unbiased accounts but because its accounts are objective enough to be proved or disproved no matter what anyone wants to be true" (D'Andrade 1995: 404). Quite simply, if we can accept some notions as "objective" (the fact that we breathe oxygen, to begin with), then we know that objectivity is not unachievable, and the various difficulties of achieving it in various (admittedly, very many) areas of knowledge should not be taken as excuse to trivialize the problem with a dismissive generalization.

2. Various forms of relativism and syncretism in post-modernist thought, and particularly the merging of different "spheres" of life (political, cultural, economic, etc.) into one single whole, are one of the main points in Fredric Jameson's *Postmodernism or, The Cultural Logic of Late Capitalism* (1991), which (especially if we consider that the seeds of his critique date back to an article he wrote in 1984) constitutes one of the earliest, most consistent and systematic critiques of post-modernism. Jameson argues against post-modernist generalizations by stating that in fact the modernist framework of analysis of those "spheres" is/was perfectly efficient, and that on the contrary the big "melting-pot" is no less than the result of capitalistic forms of colonization of the cultural discourse (Jameson 1991: 35–38). To make an example that will be employed later in one of my case-studies, the relevance of this point is particularly tangible in music, as it has resulted in several different statements concerning the claimed fallacy of genre categorizations ("there are only two types of music: good music, and bad music"—attributed to a few musicians), musicological interpretations (how many times have we witnessed the proud arrogance of a composer labeling as silly all attempts at analyzing his/her compositions?), and so forth.
3. Also, the role of the economy ("late capitalism") shall not be disregarded. If it is true, and I believe it is, that there is a connection between capitalistic marketing strategies and post-modernism, then one can only agree with Susan Sontag: "In my view, what's called postmodernism—that is, the making everything equivalent—is the perfect ideology for consumerist capitalism. It is an idea of accumulation, of preparing people for their shopping expeditions. These are not critical ideas" (http://www.iath.virginia.edu/pmc/text-only/issue.901/12.1chan.txt).
4. The simplification and relativization (which postmodernists often sell as *extreme problematicity*) of the cultural discourse in post-modernism leads to a natural trivialization of the concepts. It is what Cristopher Norris defines "a high point of sophistical ingenuity masquerading as straightforward commonsense wisdom" (Norris 1990: 5). "Trivialization", needless to say, is one of worst enemies of excellence.
5. By consequence, we witness a progressive loss of integrity and identity, in both scientific inquiry and art-making. A huge cry from the festival of creativity that modernism represented:

> […] in the dialectical leap from quantity to quality, the explosion of modern literature into a host of distinct private styles and mannerisms has been followed by a linguistic fragmentation of social life itself to the point where the norm itself is eclipsed: reduced to a neutral and reified media speech (far enough from the Utopian aspirations of the inventors of Esperanto or Basic English), which itself then becomes but one more idiolect among many. Modernist styles thereby become postmodernist codes. (Jameson 1991: 18)

The result is once again translated with the words impoverishment and trivialization, when not even loss of focus, as the following example, about parody, shows:

> In this situation parody finds itself without a vocation; it has lived, and that strange new thing pastiche slowly comes to take its place. Pastiche is, like parody, the imitation of a peculiar or unique, idiosyncratic style, the wearing of a linguistic mask, speech in a dead language. But it is a neutral practice of such mimicry, without any of parody's ulterior motives, amputated of the satiric impulse, devoid of laughter and of any conviction that alongside the abnormal tongue you have momentarily borrowed, some healthy linguistic normality still exists. Pastiche is thus blank parody, a statue with blind eyeballs. (Jameson 1991: 18)

6. A most worrying context where pastiche is manifested relates to a wider context that Jameson calls "crisis in historicity", and which has a specific application in arts:

> Nostalgia does not strike one as an altogether satisfactory word for such fascination (particularly when one thinks of the pain of a properly modernist nostalgia with a past beyond all but aesthetic retrieval), yet it directs our attention to what is a culturally far more generalized manifestation of the process in commercial art and taste, namely the so-called nostalgia film (or what the French call la mode rétro). Nostalgia films restructure the whole issue of pastiche and project it onto a collective and social level, where the desperate attempt to appropriate a missing past is now refracted through the iron law of fashion change and the emergent ideology of the generation […] Faced with these ultimate objects—our social, historical, and existential present, and the past as "referent"—the incompatibility of a postmodernist "nostalgia" art language with genuine historicity becomes dramatically apparent. The contradiction propels this mode, however, into complex and interesting new formal inventiveness; it being understood that the nostalgia film was never a matter of some old-fashioned "representation" of historical content, but instead approached the "past" through stylistic connotation, conveying "pastness" by the glossy qualities of the image, and "1930s-ness" or "1950s-ness" by the attributes of fashion. (Jameson 1991: 18–19)

I shall discuss this aspect in my first case-study (authenticity in music), but in this context it is important to add how influential the post-modernist "pastness" is in most creative industries.

7. Finally, how not to mention Alan Sokal and his spectacular intellectual joke, which said a lot, not only about the poor editorial board of the journal *Social Text*, but also about the entire post-modernist movement? Sokal was crucial in reinforcing the suspicion, from the part of many detractors, that at least some of post-modernist language is closer to bla-bla than to valuable philosophical reflection. Here is a passage from Sokal's irresistible parody:

> [natural scientists] cling to the dogma imposed by the long post-Enlightenment hegemony over the Western intellectual outlook, which can be summarized briefly as follows: that

> there exists an external world, whose properties are independent of any individual human being and indeed of humanity as a whole; that these properties are encoded in "eternal" physical laws; and that human beings can obtain reliable, albeit imperfect and tentative, knowledge of these laws by hewing to the "objective" procedures and epistemological strictures prescribed by the (so-called) scientific method. (Sokal 1996: 217)

As Sokal himself explains in the article published right after his parody (1996b), his concern over the "spread of subjectivist thinking is both intellectual and political":

> Intellectually, the problem with such doctrines is that they are false (when not simply meaningless). There is a real world; its properties are not merely social constructions; facts and evidence do matter [...] And yet, much contemporary academic theorizing consists precisely of attempts to blur these obvious truths—the utter absurdity of it all being concealed through obscure and pretentious language. [...] Incomprehensibility becomes a virtue; allusions, metaphors and puns substitute for evidence and logic. [...]
>
> Politically, I'm angered because most (though not all) of this silliness is emanating from the self-proclaimed Left. We're witnessing here a profound historical volte-face. For most of the past two centuries, the Left has been identified with science and against obscurantism [...] The recent turn of many "progressive" or "leftist" academic humanists and social scientists toward one or another form of epistemic relativism betrays this worthy heritage and undermines the already fragile prospects for progressive social critique. Theorizing about "the social construction of reality" won't help us find an effective treatment for AIDS or devise strategies for preventing global warming. (Sokal 1996b: 62–63)

Thorough analyses of the contradictions (or at least weak points) of post-modernism, as provided by those mentioned and by other scholars, does not correspond (unfortunately but predictably) to an immediate transformation in the cultural and moral values (or attitudes towards them) of the scientific and the everyday discourse. Particularly the latter (because of its size, primarily, but also because of an obviously scarcer circulation of *this type* of information) requires a longer historical process of cultural assimilation. My claim, in a nutshell, is that the influence of post-modernist ideas, and most of all *attitude*, have produced a general and diffused impoverishment/banalization of the humanist discourse and an increasing cultural laziness in our interaction with humanistic matters and topics.

In actual fact, I prefer to call it past-modernism, with an *a*. And not only (but also) in relation to Jameson's remarks on "pastness". Rather than being "post", many post-modernists seem to be mainly preoccupied with the idea that nothing can be invented anymore, and that we live in an age where illusions and media dominate. And while believing this, they practically decided to give up any form of artistic or scientific innovation, devoting themselves to the not-so-fine-anymore arts of collage, quotation*ism*, aprioristic skepticism and relativism, syncretism, applied Derridan*ism*, Lyotard*ism*, Lacan*ism* and Baudrillard*ism*, and active meta*ism*. Everything is meta-something, refers to something else and misses (in the majority of the cases) referring to itself, to have an identity, if not in a very narcissistic sense. And everything is so "aggressively inarticulate": there is an American slam poet I am a big fan of, his name is Taylor Mali, and he writes poetry of very different varieties, including some with humorous and sarcastic contents. I would like to transcribe a whole poem of his (performances of it can be easily found on YouTube, or similar platforms):

> In case you hadn't noticed,
> it has somehow become uncool
> to sound like you know what you're talking about?
> Or believe strongly in what you're saying?
> Invisible question marks and parenthetical (you know?)'s
> have been attaching themselves to the ends of our sentences?
> Even when those sentences aren't, like, questions? You know?
> Declarative sentences—so–called
> because they used to, like, DECLARE things to be true, okay,
> as opposed to other things that are, like, totally, you know, not—
> have been infected by a totally hip
> and tragically cool interrogative tone? You know?
> Like, don't think I'm uncool just because I've noticed this;
> this is just like the word on the street, you know?
> It's like what I've heard?
> I have nothing personally invested in my own opinions, okay?
> I'm just inviting you to join me in my uncertainty?
> What has happened to our conviction?
> Where are the limbs out on which we once walked?
> Have they been, like, chopped down
> with the rest of the rain forest?
> Or do we have, like, nothing to say?
> Has society become so, like, totally…
> I mean absolutely… You know?
> That we've just gotten to the point where it's just, like…
> whatever!
> And so actually our disarticulation… ness
> is just a clever sort of… thing
> to disguise the fact that we've become
> the most aggressively inarticulate generation
> to come along since…
> you know, a long, long time ago!

I entreat you, I implore you, I exhort you,
I challenge you: To speak with conviction.
To say what you believe in a manner that bespeaks
the determination with which you believe it.
Because contrary to the wisdom of the bumper sticker,
it is not enough these days to simply question authority.
You have to speak with it, too.
(Mali 2002: 37–38)[10]

"Aggressively inarticulate" is, by far, the best definition for the post-modernist *attitude* that I have ever heard. And when we couple this with the reflections we already made in the previous points of this manifesto (point 5 in particular) it becomes difficult not to agree with MacDonald (2014), however controversial she may be as political commentator, when she identifies "narcissism" and "obsession for victimhood" as two of the characteristic academic traits of our time. Plus, that narcissism is an "epidemic"—as Twenge and Campbell 2009 describes it (in fact, possibly, it is the most widespread pathology of this century)—seems to me quite evident and self-explanatory.

There is no point here, neither an intention on my part, in making a celebration of modernism and positivism, instead, but one cannot help noticing that the modernist attitude had a genuine strive for innovation, integrity and creativity, with a clear orientation towards society, rather than ego. And when we say "creativity", we mean creativity *from scratch*, not from a pre-existing model. It is significant that modernist exhibitions were abundant with words like "universal", "structure", and others that recall solidity and construction. Positivism has most certainly failed in its blindly naïve faith in progress, but a critical assessment of its exaggerations should not be translated into a *tout court* condemnation of its intentions and attitude.

To the *homo postmodernus*, on the contrary, integrity and progress are frightening: the category of truth is (not totally unjustly, by the way, but that is obviously not the point) rejected, so s/he prefers to decompose, deconstruct and take a piece from here and a piece from there. So the sexy words for today's exhibitions and installations are "fragments", "drops", "moments" and so forth. This inarticulation turns to "aggressive" when it becomes clear that it is mostly a subtle marketing strategy to expose the ego: the *you know-sort of-I mean*'s mocked by Mali are a cunning excuse to redirect the attention to one's self with the old trick of the false modesty (or, indeed, victimhood). "Look how fragile I am", "look how fragmentary a work I made", "look how insecurely I behave". Look, look, look. I, I, I. For some contents and some excellence, please come back some other time.

---

[10]I would like to thank Taylor Mali for granting the permission to transcribe his full poem in this book.

Of course, post-modernism was and remains just an example: many are the cases and the forms of scholarship that have transformed the humanities from an academic point of arrival into a point of departure.

Peirce used to say that "Doubt is an art that must be acquired with great difficulty". What we have now, in the humanities, is a loud majority of professional doubters who have made critical thought a repetitive, meaningless, narcissistic *cliché*.

The fear is: if academic research could be compared to photography, the humanities of nowadays would probably be a selfie.

Long before complaining about the inequality of resources, the hyper-bureaucracy and in general "the system", we humanists should work hard and seriously to restore excellence in our research. The hope is that projects and ideas like Numanities (and the many others that exist, starting from the ones I mentioned in point 1 of the manifesto) give a small but significant contribution in this direction.

2.8 Conclusions

It may be a good idea, before presenting some final reflection on this manifesto for Numanities, to summarize all what we have discussed so far. In most cases, I approached the various problems in a way that in Italy we call "ad ampio respiro" (something like "with a deep breath"), that is, extensively, and not sparing on digressions and incidental comments. Let us do the exact opposite, here, also to have a handier memorandum of our main points.

1. The humanities are currently facing a complex crisis that involves their impact on, and role within, society; their popularity among students and scholars; and their identity as producers and promoters of knowledge.
2. The modern western world and its economical policies have been identified as the strongest cause of such a crisis, not only *creating* the conditions for, but in fact encouraging it.
3. However, a self-critical assessment of the situation is called for. Our primary fault as humanists was that of stubbornly thinking that the world's changes could never really affect us, as—we felt—our identity was sacred.
4. The analysis of the crisis can be categorized in six main groups of approach:

    - *The "facts and figures" approach*, suggesting that the evidence of a decrease (occasionally, loss) of interest towards the humanities is documented in various quantitative ways: the drop of humanistic degrees at universities, the cut of enrollments in humanities, the restructuring of humanistic faculties (often in the direction of joining social sciences); the dramatic drop of institutional funding towards humanistic research.
    - *The "value of the humanities" approach*, suggesting that humanities should be supported "in principle" for their capacity to create "knowledge, skills, and understanding [needed] to thrive in a twenty-first-century democracy"

(Broadhead et al. 2013: 10); innovativeness, competitiveness and strength (ibid.: 11); and "leadership in an interconnected world" (ibid.: 12).

- *The "humanities need to recover strength" approach*, suggesting that while humanists keep themselves busy with "identity politics, social struggle and gender warfare" (Sheets 2013), they miss completely the great classics of arts and literature, which are challenging and demanding, missing the opportunity to understand their cultural heritage, and to engage in serious reading and critical thinking.
- *The "humanities are not really* science" *approach*, suggesting that the humanities should be studied for personal pleasure and education, and kept outside the constrictions of academic contexts.
- *The "hyper-bureaucracy is destroying humanities" approach*, suggesting that the current overload of managerial, administrative and bureaucratic tasks for academic personnel is seriously endangering their ability to operate efficiently, and to make a personal difference via their initiatives, creativity and experience.
- *The "humanities versus technology" approach*, suggesting that technology is the "great enemy" of the humanities, and a spiritual "personal self-transformation" should be the solution.

5. The main strengths of humanities are the ability to: promote critical thinking and analytical reasoning; provide knowledge and understanding of democracy and social justice; develop leadership, cultural and ethical values.
6. The main problems of humanities are the lack economic relevance; the socio-institutional perception of them as "impractical" and unemployable; the fact that they do not match with technological development.
7. The resulting crisis consists mainly in the absence (or radical reduction) of funding from institutions; a decrease in student numbers a decrease in interest; a loss of centrality in society.
8. A Numanities (New Humanities) project should consider all these aspects, with self-critical assessment on the first line. The goal is to unify the various fields, approaches and also potentials of the humanities in the context, dynamics and problems of current societies, and in an attempt to overcome the above-described crisis.
9. Discussing Numanities in terms of an "umbrella-concept" means also that there is no *specific* scientific content in it: that particularly means that the many existing new fields and research trends that are addressing the same problems (post-humanism, transhumanism, transformational humanities, etc.) are not *competitors* of Numanities, but rather possible ways to them.
10. Numanities intend to pursue a mission summarized in a seven-point manifesto.
11. First: *Rethink the position of humanities in modern society*. Regardless of the importance that they may acquire in present or future scenarios, it is crucial that humanities have a recognizable and authoritative role in research and everyday practices. To recover dignity, humanities must rethink their position and role at all levels, from the choice of research topics to the selection of the right

platform to showcase them; from the approach to writing a project application, to the whole way they *read* and *interpret* the world.

12. Second: *Reestablish the dialogue between humanities and institutions, at various levels.* Humanities should make an effort to learn the "rules of the game" of modern society. Rather than basking in self-pity for being marginalized by institutions, humanists should face the current difficulties by doing what they do best: reflect, analyze and look forward. Institutions are granting less and less funds to humanities, because they sense a lack of concreteness and profitability of the field, while humanists think that, if their research was concrete and profitable, it would simply not be humanistic. Both sides are wrong: the concept of "intellectual" has somehow split into two separate categories that do not communicate with (and in fact despite) each other: an anti-establishment "bohemian outcast" on the one hand, and the business-oriented "manager" of knowledge on the other (the latter usually working, or at least cooperating as advisor/referee, for funding institutions). What got lost in the process is the "architect" of knowledge: the intellectual who combines the vision with the engineering, the abstract with the concrete. The necessity to *learn* the "rules of the game" must be however counterbalanced by the equally-crucial necessity to *rewrite* such rules. Scholars, not just within the humanities, should take a clear stand against the overload of bureaucratic tasks.
13. Third: *Learn (or, once again, re-learn) the noble art of empirical and applied approaches.* Humanities are not able to "speak to the world" with the language of constructiveness and concreteness. Most great humanists of the past, in all fields, had the gift of connecting great visions with great actions, or—in the most forward-thinking cases—they *set the bases* for their vision to become *real* eventually. The crisis of the humanities may be also due to those many humanists who got stuck somewhere between a romantic myth and a postmodern disillusionment with their ability to actually be *rational and concrete*.
14. Fourth: *Remind humanist scholars and the rest of the world that humanities have always taken* progressive *stands within societies, not* Luddite *ones.* Humanists have been for centuries the people who *looked forward*, who were able to predict mid- and long-term orientations in history (at various levels: creative, ethical, sociological…) by means of a form of reasoning that derives from a deep understanding of the mechanisms and actions of a society, both at small and large scales. Several humanists are now becoming *uncritical scholars of the past*, and occupy the scholarly niches of traditions, folklore and philology, with an exclusive interest in memory, conservation and identity. The crucial contact with *values* is occasionally endangered: failing to keep up-to-date with a society that is clearly not placing humanities at the center of attention, may resolve into pub-like reminiscences of the past, rather than a proper study of it. The bound with reality is lost and the "unknown" makes us fearful and lazy. The point in Numanities is not to accept new technologies in principle, but rather to give a proper, careful scrutiny of them.

15. Fifth: *More specifically, in the current context of economic, political and ecological struggles, to be progressive should mean to be ethically-minded, sustainable, quality of life-oriented and dignity of life-based*. In the words of Gandhi first and Schumacher later, progress is implemented through *appropriate technologies*. "Progress" is not necessarily represented by every single thing that is newer, bigger, faster: a critical (therefore, intimately humanistic) view seeks progress in what is tailored for the real needs and demands of a given community. Humanities have the theoretical skills to offer a constructive and critical view on these matters: Numanities would like to reestablish this interest and additionally cultivate operative and concrete tools so that those skills are not wasted in purely abstract, distant-from-reality formulations.
16. Sixth: *Promote, with pride and determination, those values that the other fields of knowledge are simply not capable of formulating*. If natural sciences study and create what makes life *possible*, humanities study and create what makes life *worthwhile*. Creativity, beauty, quality of life, dignity of life, empathy, tolerance, culture, reflection… this is what humanities do best, and what humanities are mostly for. If society and institutions have become reluctant to understand the importance of these values, then humanities need to find new ways to convey the message. The crisis of humanities involves also their identity as producers and *promoters* of knowledge. Indeed, part of the enormous heritage that humanities have delivered to the world consists also in constantly stressing the importance of knowledge and culture. They did not just provide material to acquire them: they were also the main PR of them. That is nowadays more important than ever, as the need for knowledge and culture has become increasingly marginal in modern society.
17. Seventh: *Strive for the excellence of humanities*. The humanities have to restore rigorous standards of academic excellence. Changes in society aside, *this* is one of the prominent factors that are giving humanities a bad reputation: the *noise* of mediocre research has increased in volume, and therefore it is more and more difficult for institutions to spot excellence (see Fig. 2.4 for a recapitulation of the manifesto).

In the light of these premises and reflections, it is now possible to identify the areas of inquiry that Numanities, in their functions and comprehensive approach, seek to cover. The following list should also be understood as a statement of purposes for this entire book series. These, in other words, will be the topics/areas we intend to represent (see Fig. 2.5 for a summary in bullet points).

In no particular order:

(a) **Traditional fields of humanities**—as long as their research paths are focused on issues of current concern (e.g., art research that addresses questions of the redefinition of art in the light of modern technologies). This statement may easily be misinterpreted as disrespectful to the history of the humanities, so it may be useful to reiterate that we are not talking about what deserves or not to be part of humanistic research nowadays. It is rather what *Numanities* (with an

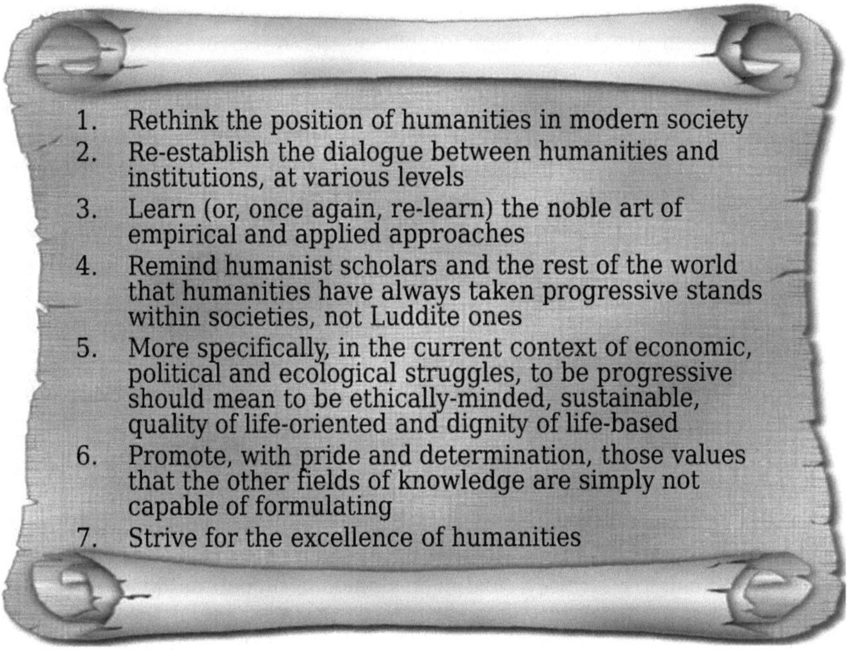

Fig. 2.4 The manifesto of Numanities in seven points

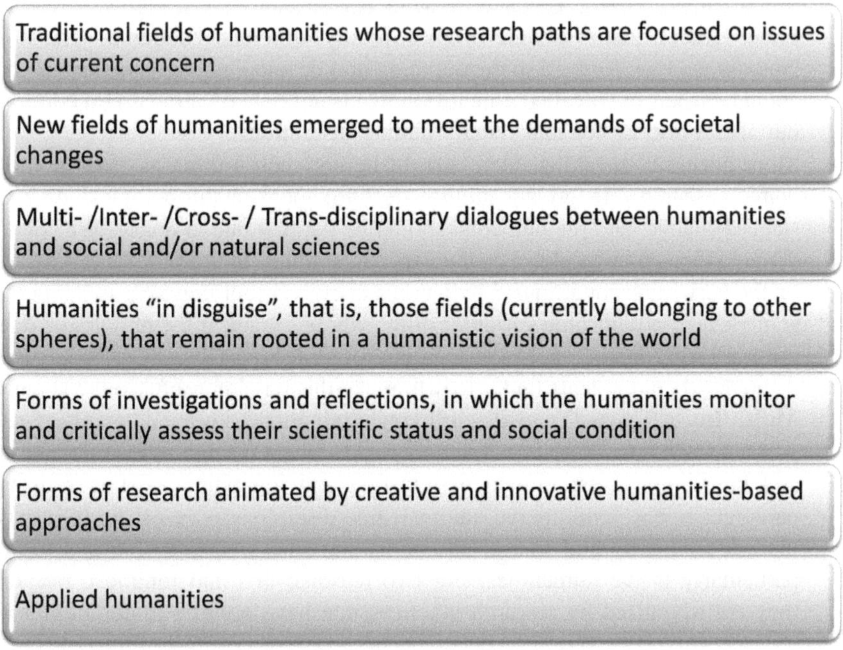

Fig. 2.5 Topics and areas covered by Numanities

N) would like to pursue and host. There is a threshold, perhaps a thin one, above which a "classic" form of humanistic research is pertinent to the Numanities' platform, and below which it is not. The determination of such a threshold lies in the relevance of a piece of given research (even one about ancient events or characters) within the context of current needs and problems in society. Whenever these are addressed in an appropriate and tangible manner, a humanistic inquiry is also a numanistic one. "Appropriate and tangible", to be clear, means that explanations such as "this research is relevant in principle" or "this research is relevant because it increases our knowledge on that particular topic", alone, are not enough to pass the threshold. What it is proposed here is *not* the removal, from the humanities' universe, of—say—an obsessed Italian philologist who can devote his/her entire life to digging the world's archives in search of every single document of Lodovico Castelvetro's and Annibale Caro's heated debate over one of the latter's poems in the 16th century: what is instead being said is that it is very difficult to place such works into the development of Numanities, or at least we have a very hard time in understanding how on earth this kind of enterprises will help us understand and improve our society. And—to make matters more cynical—it may be even more difficult to convince a funding institution to support this kind of research.

(b) **New fields** have recently emerged to meet the demands of societal changes (e.g., digital humanities). On the other hand, to keep up with the same example, if Castelvetro's and Caro's polemic is "rejuvenated" within—let us say—a project of digitalization of 16th century documents, in order to make them available to a larger group of readers and accessible to disadvantaged social categories (for instance, a blind person can have a software reading these letters), then we are into a totally different ballgame—one that Numanities look forward to encouraging and developing. Now we are talking about a project of so-called e-inclusion that is not only of very current concern, but also a priority for many state funding institutions.

More generally speaking, there should be not such a thing as "frozen" academic research. Historical, cultural, technological, anthropological circumstances change, and the humanist should be the first one to notice. We cannot demand to keep on working on the very same things, in the very same way and on the very same conditions. Mediation and even compromise has to happen at some level, and on some variables of the whole setting. If the philologist does not want to "bargain" on his/her passion for Castelvetro's and Caro's *tenzone*, then s/he should at least find *a way* to make this passion useful to the society.

(c) **Multi-/Inter-/Cross-/Transdisciplinary dialogues** between humanities and social and/or natural sciences (namely: fields where disciplinary combinations and interactions were crucial in updating and upgrading existing research paradigms—e.g., multimodal studies). Interdisciplinarity, crossdisciplinarity, multidisciplinarity and transdisciplinarity have become rather abused terms, particularly in the humanities. For two reasons: first and foremost, because they are trivialized by research practices that have nothing in common with any of the four methodological procedures; and second because they are by now employed as synonyms, while they represent four tangibly-distinguished strategies (with "inter" and "multi" being the most recurrent prefixes, and therefore the more easily confused approaches). The humanist adores sounding modern and relevant by calling his/her research "interdisciplinary": most of the times, however, what we witness is far from any real synergy across, and therefore challenge to, disciplinary boundaries, but rather an ordinary reiteration of an existing convergence between two fields that have been cooperating for ages, already. It is time to state clearly that there is no "multi-inter-cross-trans-" achievement in, say, a literary theorist who needs some concept from political sciences to analyze the work of Pablo Neruda. Concepts, words and definitions are not unlike the art of cooking. The moment you add one ingredient after another, in the illusion that *the more you put, the better it tastes*, is the moment when you are killing a dish. The dialogue/confrontation among fields of knowledge is a serious activity that cannot be oversimplified with a simple "exchange of information" between two fields that are anyway inclined to act thus.

In other words, Numanities intend to encourage (1) *real*, *consistent* and *challenging* dialogues between the humanities and other fields; and (2) a proper distinction, and consequent appropriate choice of methodological strategies. A multidisciplinary piece of research is not an interdisciplinary one, which is not a transdisciplinary one, etc. Just as a reminder: **crossdisciplinarity** consists of one discipline approaching a research problem from the perspective of another discipline; **multidisciplinarity** consists of the cooperation among different disciplines, each drawing on their intradisciplinary knowledge; **interdisciplinarity** consists of an organic integration/synthesis of theories and methodologies from different disciplines; and finally **transdisciplinarity** consists of the creation of a new disciplinary unit, as generated by the coherent junction of different disciplines.

(d) **Humanities "in disguise"**, that is, those fields (currently belonging to other spheres, particularly social sciences), whose "spirit", attitude and methodologies remain rooted in a humanistic vision of the world (e.g., soft power theories, future studies, etc.). An element of pride, admittedly, is hinted here. Generalizing a bit, the recent history of the humanities displayed a peculiar form of diaspora. On the one hand, some research areas remained anchored to the humanities, in terms of both identity and social-institutional perception, becoming the main "victims" of the crisis. On the other hand, other areas managed to "migrate" towards more prosperous shores (social sciences, mainly, but not only), often surviving the crisis, or even, in the luckiest cases, reinforcing their position within academia. The migration process was so strong, that many of these areas disappeared from the humanistic galaxy and reappeared in another.

The irritation, and perhaps the jealousy, of the former group towards the latter is very apparent through writings, conferences and the likes: the "victims" accuse the "migrants" of betraying their identity, losing integrity, and calling the same things with different, more appealing, names (Nelson-Gaonkar 1996: 28–29, describe for instance how the rise of the "trendy" cultural studies have angered at least the communities of anthropologists and literary scholars). The suggestion, here, is to make sure that fields deriving from humanities, or humanistic in practice (whether or not willing to admit it) get involved in the numanistic discussion as, so to speak, "full members" of the community. To take but one example, as a semiotician I was always a bit disappointed to see that a discipline that is so clearly embedded in signification, codes and sign processes such as future studies had successfully managed to avoid any official connection with semiotics, with the tragic exception of a study program in "Semiotics and Future Studies" arranged by the George Fox Evangelical Seminary, and advertised for "leaders adept at seeing signs of Jesus' work in the world–leaders who can proactively guide the church into the future of this era of uncertainty and change" (http://www.georgefox.edu/seminary/programs/dmin/sfs/). It was therefore a very pleasant surprise, on the occasion of the first International Congress of Numanities organized by the International Semiotics Institute in 2014, to receive the submission for a whole session in Future Studies directed by Osmo Kuusi (Aalto University, Helsinki) and Toni Ahlqvist (VTT Technical Research Centre of Finland). Dialogues of this type, too, are possible.

(e) **Meta-humanities**, that is, forms of investigations and reflections, in which the humanities (or any of their affiliated disciplines), monitor and critically assess their status, paradigms, scientific and social condition, impact, and so forth. It is necessary, perhaps even obvious, that a significant role in this forum is played by thorough discussions on the state of the art of the humanities and on the possible next moves. The whole Numanities project, for one, is primarily a meta-humanistic enterprise, at least in its foundations. However, this necessity should not be mistaken for an "emergency action" in times of crisis. Just as a more balanced diet is not only a temporary solution to some health problem, but should also remain as a lifestyle, a wise habit of self-analysis should accompany humanists not only in difficult times, but also—perhaps mostly—when these times are overcome. Once again, the need for more humility and less narcissism comes to mind, here.

(f) Forms of research animated by **creative and innovative humanities-based methodological approaches** (for instance, the increasing trend of adopting artistic methods in natural sciences). In other words: forms of research where the humanities play the role of the "added value". During Stanford University's 2011 BiblioTech conference on "Human Experience", Damon Horowitz, director of engineering at Google, famously delivered a talk entitled "Why you should quit your technology job and get a Ph.D. in the humanities":

> In our Internet-enabled era, it is easy for technologists to parlay creative power into societal power: We build systems that ease the transactions of everyday life, and earn social validation that we are "making the world a better place." Within a few years I had achieved more worldly success than previous generations could have imagined. [...]
>
> But there was a problem. Over time, it became increasingly hard to ignore the fact that the artificial intelligence systems I was building were not actually that intelligent. They could perform well on specific tasks; but they were unable to function when anything changed in their environment. I realized that, while I had set out in AI to build a better thinker, all I had really done was to create a bunch of clever toys—toys that were certainly not up to the task of being our intellectual surrogates. [...]
>
> I wanted to better understand what it was about how we were defining intelligence that was leading us astray [...] and, slowly, I realized that the questions I was asking were philosophical questions—about the nature of thought, the structure of language, the grounds of meaning.
>
> [...] Thus, about a decade ago, I quit my technology job to get a Ph.D. in philosophy. And that was one of the best decisions I ever made.
>
> [...] As I learned about those things, I realized just how limited my technologist view of thought and language was. I learned how the quantifiable, individualistic, a historical—that is, computational—view I had of cognition failed to account for whole expanses of cognitive experience [...] Most striking, I learned that there were historical precedents for exactly the sort of logical oversimplifications that characterized my AI work. Indeed, there were even precedents for my motivation in embarking on such work in the first place. [...] In learning the limits of my technologist worldview, I didn't just get a few handy ideas about how to build better AI systems. My studies opened up a new outlook on the world. I would unapologetically characterize it as a personal intellectual transformation: a renewed appreciation for the elements of life that are not scientifically understood or technologically engineered.
>
> In other words: I became a humanist.
>
> And having a more humanistic sensibility has made me a much better technologist than I was before. (Horowitz 2011)

While in point 6 of the manifesto I possibly showed a less optimistic side of the story, claiming that the humanities have to return to "add value" to society, in order to be "added values" (implying, thus, that they have mostly mislaid that skill), passionate endorsements like Horowitz's talk, show hope and confidence for the immediate future, and—what is particularly significant—acknowledgement from the holders of "societal power".

Another example on the same line, but in a totally different context comes from zoomusicology, a field I have been dealing with quite extensively in the past (see, for instance, Martinelli 2002 and 2009): it is the case of the Senegalese percussionist Arona N'Daye Rose, who enjoyed a certain popularity and gratitude among marine biologists for having decoded an apparently incomprehensible recording of cachalots (who notoriously communicate by means of short clicking sounds) for the benefit of the scholars André and Kamminga (2000). What seemed to be an absolute mess of clicks for the two scientists, was simply an inventive, yet orderly, rhythmic sequence for Rose, who, taking a musician's stand, found it very easy to detect how many cachalots were vocalizing in the recording.

Finally, later on in this book, as we will discuss the development of Numanities within the International Semiotics Institute, I will also mention a couple of successful research projects implemented by one of the institute's researchers, Mindaugas Gapševičius, which are exactly centered around the difference that arts can make within natural sciences.

Making a difference and adding value are two tasks that the humanities can perform very competently, if they accept putting themselves at stake and embracing the changing world critically but without prejudice.

(g) **Applied humanities**. This point is self-explanatory: we need more and more situations in which the humanities are able to go beyond the academic territory and put their research into effect in everyday life and matters. It is quite interesting that the most popular online source of information, Wikipedia, in a specific entry called "Outline of the Humanities", considers only three areas for "applied humanities":

> Applied arts such as fashion design, graphic design, industrial design, interior design
>
> Health humanities—application of humanities disciplines to discourse about, expression of, and/or the promotion of the dimensions of human health and well being.
>
> Medical humanities—is an interdisciplinary field of medicine which includes the humanities and their application to medical education and practice.
>
> (https://en.wikipedia.org/wiki/Outline_of_the_humanities)

Can this really be the end of the story? Well, to begin with, Wikipedia seems to have forgotten the most obvious forms of applied humanities: journalism, creative writing and anything related to cultural heritage and conservation (libraries, museums, exhibitions…). Those, along with the listed "applied arts", are the classics of the category, so to speak. But there is more. In recent years, several humanities programs with an applied orientation, particularly in philosophy and history, have been established in various institutions. Degrees in applied humanities can be pursued in many universities: Californian institutions seem to have a particular inclination for them (Claremont, Brandman, Davis…), but programs can be found everywhere in the world, from Carlow Institute of Technology, Ireland, to the University of Canberra, Australia. Plus: places like Georgia State University have programs in applied philosophy; Carnegie-Mellon in Pittsburgh and the University of California at Santa Barbara offer degrees in applied history; Brown University, Rhode Island, has a program in public humanities, etc.

These programs are not only forming journalists or graphic designers: they are creating new professional figures that reflect the importance of engagement with the wider community at local, national and international level. They prepare students to become creators, protectors, and advocates of art and cultures with far-reaching impacts on more than one level. Applied historians, for instance, work in seeking clues to contemporary problems in the records of earlier generations coping with similar problems. Such figures can enter governments, consulting firms or again academia. Applied humanists, with learning outcomes dedicated to modern technologies and communications, can work in digital media, digital marketing and the likes. In sum: there is a whole world, outside universities, where humanist professional profiles are useful and desirable (not to mention the acknowledgment of them as something that can "make a difference", as we saw in the previous point, through the words of Damon Horowitz): it may only be my personal impression—so I may be very wrong here, but it seems to me that before the crisis humanists tended to consider extra-academic jobs as second-rate ones, while during the crisis they failed to see them as a possible solution to it. The area of "applied humanities", to my mind, must be (1) ennobled a bit, particularly in its intellectual value (again, it is only my impression, but the feeling is that the main reason why humanists look down to these jobs is because they secretly think that those who do them, do them because they are not smart enough to be academics); and (2) better understood and analyzed, particularly in its recent developments that brought the creation of new professional figures.

Discussions of this kind, too, are welcome in the Numanities.

In the light of all this, keeping in mind the analysis of the humanities' crisis, our 7-point manifesto of intents, and the attempt to place the Numanities' project into concrete areas of inquiry, I would like to conclude this part with the definition of three pillars for Numanities:

1. The **Epistemology** of Numanities goes through a basic form of validation: making a difference, being an added value. Numanities promote forms of research where the humanities are not only precious in the economy of a whole project, but can add something unique that other fields are not capable of delivering.
2. The **Methodology** of Numanities is embedded into a true interaction with other fields, particularly the natural sciences and technologies. Depending on the case, humanities can embark into multi-, cross-, trans- or interdisciplinary enterprises (and they know the difference between them).
3. The **Impact** of Numanities is measured on their capacity to work for/from/through/in society. They cannot be an end in themselves. For this reason they have to be R.E.A.L.: reasonable, empirical, applicable and logical.

The time has come for us to prove how excellent, determined, sustainable, appropriate and concrete humanities are capable of being. We will need good ideas, which is fine, because humanities are about ideas; we will need to learn the new languages of modern society, which is still fine, because the study of languages is a field of humanities; and we will need to be rational, humble and practical, which is the only down-side, as these qualities *used to be* typical of the humanities, but we lost a bit contact with them.

Chapter 3
Case Studies

Abstract This part of the book will present two case studies based on two relatively recent (and now updated) investigations that I performed in the fields of musical semiotics and zoosemiotics respectively. These particular two were chosen as they were considered especially relevant within the discussion of Numanities, and in two different ways. In the first case study I discuss the most typical result (or perhaps cause?) of the technophobic attitude that I have addressed very often in this book: the issue of "authenticity", as applied, in my particular study, to music (and even more specifically to popular music). In the second case study, I analyze two different forms of comparative analysis between human and non-human cognition: The question of human specificity within the animal kingdom, and the question of language and interspecific communication. Like in the former case, this study, too, is aimed at a critical commentary on (what I consider) redundant biases in current humanistic research—anthropocentrism and speciesism.

3.1 Introduction: The Role of Semiotics

Before I inaugurate these two sections, the time has come to be a bit clearer on the role of semiotics within Numanities. Readers must have noticed the recurrence of this term throughout the text, and will soon notice the centrality of this discipline in my two case studies. Most of all, it must have not escaped that the whole project was generated within an academic unit called the International Semiotics Institute (and will read a lot about this process in the appendix to this book). So, a few questions must have arisen in the readers' minds: why would an idea like Numanities be produced within a semiotic context? Is semiotics "numanistic" in principle? Is it "especially" numanistic? And why?

To begin in the most trivial of ways, semiotics is a field within the area of the humanities (most of the times, at least), so a discourse on "new humanities" is a discourse that concerns semiotics as well. Obviously, though, this is not enough of a reason to claim a leading position in a paradigm which is clearly intended for each and every discipline of such an area (and in fact, as I shall soon specify, there is no

D. Martinelli, *Arts and Humanities in Progress*, Numanities - Arts and Humanities in Progress 1, DOI 10.1007/978-3-319-45553-2_3

intention to assign a "leading position" at all). A second trivial reason for this enterprise is merely institutional: if we consider that the project was started (a) at the International *Semiotics* Institute (as we just said), (b) within the Faculty of Social Sciences, Arts and *Humanities*, and (c) at Kaunas University of *Technology*, it almost seems natural that this interdisciplinary set of Chinese boxes would produce a notion of that kind.

However, there are more pinpointed reflections that I would like to share that concern (what I perceive to be) the relevance of semiotics within this novel theoretical enterprise. A disclaimer, first: as it has already been for a few years (see Martinelli 2009) that I have been preaching more humility among semioticians (particularly the necessity to soften the pretention that semiotics is the best tool to explain every single phenomenon on this planet—and on others too), it should not be surprising that, even in this case, I will not claim that semiotics is the only or the best paradigm for embracing an enterprise like the one above-described. At the same time, I feel there are some promising preconditions for semiotics to play a significant role in the discussion, without necessarily take the leading one.

First and foremost, by its own nature, the field can be considered in all respects a part of Numanities. Thomas Sebeok once said that semiotics is "the name that philosophy took during the 20th century", and while that may sound a bit bold (and it probably is), there is an element of truth in granting semiotics with the merit to have brought communication and mediation to the center of the philosophical discussion, in a time, the last century, when indeed these two phenomena had become central in general. Moreover, semiotics was a forerunner of truly interdisciplinary studies. Interdisciplinarity was not an easy task during decades, such as the 1960s and the 1970s, when academic programs and research strategies were rather strict and monolithic, yet semiotics managed to produce a refreshing type of intellectual, who would be able to connect the dots of several disciplines through a common methodology. More significantly, that attitude eventually transcended the boundaries of humanities, and started involving more and more areas of natural sciences (zoosemiotics, cybersemiotics, etc.), becoming a pole of attraction for many scholars from the most diverse fields. Of course, in more recent years, this whole enterprise became the main source of some semioticians' superficiality and arrogance (ending up alienating many of those sympathies from other fields): however, if we keep the focus on the potentials of the discipline and on its best calibrated applications, what we get is a versatile field that can provide valuable tools for understanding society in the way that we have described in the first part of the book.

A confirmation of this, and at the same time another important argument, is the fact that some recently-born or recently-developed fields of inquiry have incorporated semiotic methodologies (and their "original spirit") in their paradigm, sometimes admitting it (as in the case of multimodal studies), sometimes not quite (as in the case of cultural studies). In any case, acknowledgment is not the point; the point is that, contrary to an increasingly widespread opinion, semiotics is not agonizing: it is rather recovering its original identity as a "methodology", rather than a "complete and autonomous theory". And, if a slightly polemical verve is allowed, once some of the most valuable semioticians—those who instead are now engaged in megalomaniac enterprises of semiotic interpretation of the universe—will realize this, times for a new spring of semiotics will be mature. For the record, I shall elaborate on this "megalomania" in my second case study.

A third argument: although in different shapes and degrees, semioticians have always kept an eye open on ethical and political issues, managing (most of the times successfully) to compromise scientific impartiality with socially-committed statements. This is an important aspect, because it keeps the field in close (and concrete) contact with reality as such, not limiting its activity to the observation of it, but becoming proactive about it (and that, as we have seen, is a primary task for Numanities). This was not easy to accomplish, as similar enterprises are always regarded with enormous suspicion by those who think that science should be ethically neutral—as if this (a) was possible, and (b) could be distinguished from being "ethically numb", as certain experimental research has proved in many cases.

Predictably, semioticians themselves have been discussing the limits and the potentials of such moves. Most of all: does semiotics, the way it is constructed as a field of inquiry, have in principle the epistemological prerequisites to include ethics in its theoretical program (for an interesting insight into this discussion, see Bopry 2002: 15–16)? Of course, even readers who are not deeply familiar with semiotics may recall at least the cases of Roland Barthes' *Mythologies* (1957), and possibly Ferruccio Rossi-Landi's works on language and ideology (such as 1972, or the entire project of the journal *Ideologie*, which he himself founded). In addition to these, there is an entire (lesser known to non-specialist readers) tradition of specific applications of the semiotic paradigm to ethical issues, up to straight identifications of semiotics with ethics (it is the case with Ponzio and Petrilli's project of semioethics). Among these applications, one should at least mention Stefani (1985, 1989) (where musical semiotics is put in connection with peace studies), Stefani and Guerra (2005) (where several issues in social integration, disabilities and art-therapy are addressed through a semiotic interface), Habermas (1998) (on communicative ethics), Tarasti (1997, 2000) (on the general necessity of ethically-minded research in semiotics), and that interesting path on the semiotics of resistance that goes from Eco (1997) to Bankov (2004), then to Tarasti (2005), and (on a specific zoosemiotic application) to Martinelli (2008). Particularly in 2000 Tarasti makes the claim that a semiotician does not have to fear dealing with ethical issues, as applied to semiotic research. As he himself states about art:

> [W]e cannot exclude from the present investigation the vast domain of ethical problems. One cannot be satisfied, in general, with a study in which a phenomenon is reduced to some of its aspects, but even in a most abstract art work we have to account for the whole weight of the reality that has yielded it and which speaks therein through its own sign systems. The intentions of an author cannot be eliminated as a kind of intentional fallacy. The social context and the "ecoform" of an art work cannot be left without attention as "extrasemiotic conditions". (Tarasti 2000: 87–88)

In what he calls the "discovery of ethics", Tarasti underlines the importance of the subject and its foregrounding and disappearance: "Without the concept of subject there is no ethical choice" (Tarasti 2000: 88). It seems, thus, that a prescriptive semiotics, as an implement to a descriptive one, does not sound anymore like an anti-scientific semiotics, as it may have done not too long ago, when it was suggested that science should prevent personal opinions from emerging, and—more to the point—that suggesting "what people should do" is not so scientific as "explaining phenomena". The cases mentioned above, and several others which are emerging in the young generation of semioticians, show the scholars' preference for an intellectually honest weakness to a pretended impartiality, which—it is argued—is by principle impossible.

Back to Numanities, the golden rule, here, is: providing a commentary on society and reality is, to a certain extent, a moral duty for humanists. If what we mentioned in the manifesto is true, that is, that humanities' object of study is "making life worthwhile", then—once again—contributing to the reinforcement of values like culture, quality of life, tolerance, and so forth, cannot be anything else than an ethical statement. Let us not be afraid of that: there are no ghosts of authoritative regimes' propaganda, here. It is an entirely democratic enterprise that simply takes to the letter the necessity of creating a "value-based" society.

One more argument in favor of matching semiotics with Numanities is the fact that semiotics is a guarantee of support for conservative scholars, who (after all I have stated in the first part of the book) might regard this project with a rather suspicious attitude. Numanities, as already remarked, are in principle embracing an "appropriate technologies" philosophy: there is no progress or regress as such, but there are strategies that are tailored to the needs of communities and single individuals. As we said: neither forwards nor backwards, but towards. The problem is that an approach like this is already more progressive than conservative, because it represents a change of mentality in a direction that most people will not be familiar with. This, of course, is already enough of a reason for a conservative-minded scholar not to be attracted by a similar perspective. However, what appropriate technologies actually reject is *unmotivated* regress, just as well as unmotivated progress. Whenever reasonable arguments to take one direction or another are brought in, appropriate technologies (and therefore Numanities) are ready to listen with great attention. In fact, quite a few times, in this book, we have looked at the humanities' "past" as inspiration for the future: in all those instances, we found that the "old" identity/role/paradigm of humanities was more sustainable and "appropriate" than the present ones.

Now: it is intriguing that semiotics, especially in the last 20–30 years, has become more bipartisan than it used to be in earlier days, when progressive, socialist and occasionally revolutionary figures like Eco, Barthes, Rossi-Landi, Greimas, Foucault, Stefani, Tagg and many others were dominating the scene. The development of other currents, particularly the Peircean one from the 1980s onwards, has instead brought to prominence scholars who are rather ideologically-disengaged (which already is not the most socialist of moves), neoliberal, or indeed of a more conservative extraction (though we have no straight fascists yet—at least I hope so). As a consequence, not all semioticians engaging in ethically-minded reflections have

Marx or Lenin as points of reference: a number of vigorous and consistent statements came also from the "right" side of the political spectrum, promoting (as in the case of the "Semiotics of resistance" path) ideas and ideals of conservation and respect of a past that—it was argued—was more suitable than the present to answer some important demands of society and academic research in particular. There was no need for the scholars involved to be politically aligned to a right-wing party or ideology, as the point was the ideas as such: the semiotics of resistance departed from Umberto Eco (someone who certainly does not mind being associated to a left-wing philosophy), and then continued with Kristian Bankov (who equally does not mind being associated to the right) and Eero Tarasti (who instead would probably define himself as politically neutral). The good news is that those statements were by all means "appropriate technologies", in that concepts and visions were presented in a theoretically accurate and motivated manner, resisting the temptation of simplistic "Oh, those were the days!" affirmations.

To agree or not with such ideas is of course not important, in this sense: what we know is that semiotics has proved its ability to produce pondered and grounded ethical reflections, regardless of their ideological connection, ensuring an impartial platform for different sorts of discussion.

Last, but definitely not least, we have the theoretical mission of semiotics: the study of signs, codes, communication, representation, and so forth. I would find it extremely difficult and peculiar to engage in a thorough analysis of modern society, its demands and its perspectives, without having a proper set of tools to understand the dynamics and the signification of the various phenomena involved. I previously mentioned future studies as one of the "humanities in disguise": the field has gained an increasingly positive (and deserved) reputation, managing to dismiss the prejudice of those who saw the field as a kind of "astrology". Now—guess what—future studies are literally packed with semiotic notions, starting from the field's basic assumption that if you want to notice in the present the symptoms of what may actually happen in the future, you must have a theory of signals. Which is indeed what future studies have. Putting the issue of humility aside for a second, I cannot help remarking that the best and most appropriate tool for producing a theory of signals is indeed semiotics.

3.2 Case Study 1: Authenticity and Popular Music

The first case study I shall like to address in this section of the book has a lot to do with the problems of technophobia and "academic Luddism" that I emphasized in different parts of this book. The issue of authenticity has become a remarkably fashionable (redundant?) occurrence in both scholarly and everyday discourses about the arts (not only about the arts, in fact, but mostly about them), and—to an extent—it has turned into one of the defense fortresses (along with the concepts of "tradition", "identity" and a few others) that a few humanists employ to resist the advancement of technological innovations, or—in the most extreme cases—"change" as such (in research as well as in society). Having devoted some research to the issue of authenticity in popular music (see at least Martinelli 2008, 2010b), I shall like to build on those reflections, adding more emphasis on those aspects that are more sensible to the numanistic discourse we are developing here.

In Shuker's study (1998: 17), which specifically deals with Popular Music, the concept is distinguished according to three forms of usage: a commonsensical one, referred to the elements of creativity and originality of the musical work; a marketing one, addressing the (very popular, as we shall see) dichotomy authentic-commercial, where the two features are in an inversely proportional relation; and a (sub)cultural one, where given artists or works are "legitimized" within a community, specifically in relation to the degree of musical authenticity exhibited.

Grossberg (1992: 208) adds a specific stylistic extension to the concept, with such oppositions as live performances versus club culture (with the former obviously representing the authenticity status, *in opposition* to the latter).

In these treatises, and others too (Thornton 1995; Moore 2002 being two further notable examples), the common denominator seemingly exists in the argument that the label "authentic", as applied to works, composers, styles, performances or other things, bears a clearly positive ideological connotation (both in the epistemological and, as we shall see, in the political sense), as if it was a feature to pursue almost systematically, paradigmatically, when producing an artwork. As Peter Kivy, somewhat ironically, suggests about musical performance:

> The highest praise one can bestow on a musical performance, in many influential circles, is to say that it was "authentic" … "Authentic", then, has become or is close to becoming a synonym for "good", while seeming to confer upon a performance some magical property that it did not have before. It is the musical version of the doctrine of the real presence. (Kivy 1995: 1)

Apparently, the qualities associated with authenticity are usually of an ethical type (which is mostly why they embody some of the conservative humanists' fears): authentic music is perceived as honest, real, uncorrupted, and even "loyal" towards the audience. The fear is inauthenticity: the kitsch, the fake, the too artefact, too commercial, too technological, etc.

Many are the musicologists who addressed the issue of authenticity in explicitly ethical terms, in fact reminding us of the importance (which I by all means support) of forms of scientific inquiry that are not only descriptive, but accept the challenge of being prescriptive as well. In the introduction to this chapter, for instance, we mentioned Eero Tarasti's reflections on ethics (2000: 87–88): quite significantly, he applies them exactly on a chapter of his book named "On the Authenticity and Inauthenticity of Art". For the purposes of my analysis, that chapter provides an excellent theoretical hook to reflect upon these topics.

3.2.1 *Contextualizing Authenticity*

Following Umberto Eco,

> The necessary conditions for a forgery are that, given the actual or supposed existence of an object Oa, made by A (be it a human author or whatever) under specific historical circumstances t1, there is a different object Ob, made by B (be it a human author or whatever) under circumstances t2, which under a certain description displays strong similarities with Oa (or with a traditional image of Oa). The sufficient condition for a forgery is that it is claimed by some Claimant that Ob is indiscernibly identical with Oa. (Eco 1987: 9)

In other words, a "case" for authenticity is given when an example of inauthenticity (a forgery, in Eco's words) is produced (see also Fig. 3.1). That is to say, Oa becomes recognizable through the emergence of Ob. This is a rather important (and, as we shall see, controversial) point that certainly deserves further reflections (which I shall provide later). At this early stage of my analysis, however, I shall focus on that variable "t" of Eco's model, the historical circumstances. A historical circumstance goes certainly beyond a strictly temporal context (Eco does not say it explicitly, but I have no reason to doubt he implies it), and refers, I believe, to a rather wide range of contexts. Or, at least, so we have to assume, otherwise the level of superficiality of a model including an exclusively temporal variable would be suspiciously high. A "circumstance" can be evidently also geographical, social, anthropological, and so on. A unitary interpretation could still be possible if changes within the variable occurred coherently, i.e., if a change in one sub-unit of "t" (e.g., the geographical one) provoked an analogous change in the other sub-units (historical, social, etc.). However, this is definitely not the case. Cover versions of a song can be issued in a different time, but in the same place as the original, or—more curiously—vice versa, as it was very common in Italy with Anglo-American songs during the 1960s.[1]

To say it all, we also have to assume that Eco implied that the subjects A and B are not invariably two different subjects, but in principle they could be the same subject in two different circumstances. The production of the token (i.e., forgery), indeed, does not necessarily require a subject B, that is different from A. B may be a sufficient, but not a necessary condition for a forgery to be produced. The cases of live performances in popular music are prototypical. During their promotional tours, musicians perform several forgeries of their own material, in many cases

[1]It was a rather peculiar phenomenon due to two factors: (a) unlike today, album releases did not occur at the same time everywhere (the first album to be released on the same day all over the world was notoriously The Beatles' *Sgt. Pepper's Lonely Hearts Club Band*, in 1967), and Italian fans could literally wait months before an English or American album would appear in the peninsula's record shops; (b) the Italian copyright society SIAE had back then the curious rule of considering lyric translators at the same level of lyric authors, in terms of authorship acknowledgment and, consequently, royalties. The amusing result of this status quo was, thus, that an Italian cover of a popular foreign hit (e.g., "Stand by Me" which became "Pregherò") could actually be released in Italy *before* the original and credit the Italian translator of the lyrics along the real authors of the song.

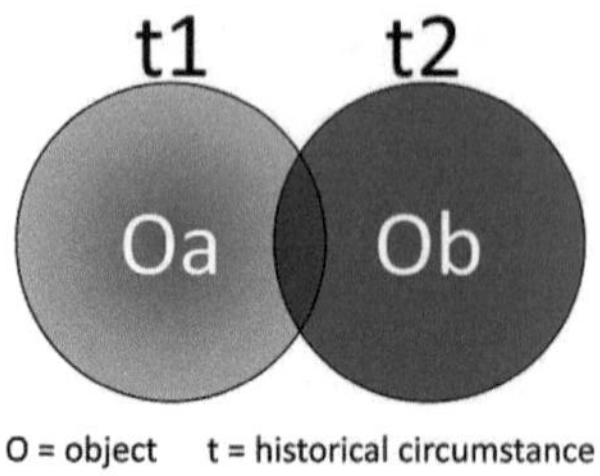

Fig. 3.1 The emergence of a case of "forgery" according to Eco (1987)

transforming the songs in length, arrangement, form, harmony, rhythm and even melody, up to creating sheer stylistic clichés around the live version. The very famous riff of Deep Purple's "Smoke on the water", in the re-made live version released on the "Live in Japan" album (the riff is totally different from bar 5 to bar 11), is not less famous than the studio-recorded version. The same applies to the habit, nowadays almost praxis, of re-issuing old material with new arrangements, or even only with a different (usually digital) mastering of the original (usually analog) tape.

Coming back to the variable "t", it is evident that authenticity and inauthenticity are features that apply at the same time to diachronic and synchronic categories, and therefore the "contexts", Eco's historical circumstances, range from the simplest and most obvious classifications to the most specific and pinpointed ones. Here is a brief list (I shall take the liberty to elaborate a bit on some of them):

Geography. The most predictable circumstance for this dualism regards the birthplace of a given type of music (form/tradition/genre/opus etc.), i.e., in most cases, the place where the given element was first found. In popular music, we are surrounded by "authentic Philadelphia soul", "authentic Nashville country", "authentic Argentinian tangos". The fight of Finnish people to emancipate their own tango tradition to the eyes of the world is undeniably a fight for authenticity. Tango was admittedly imported from Argentina and Uruguay, *but*, Finns love to repeat, by the end of the 1930s the Finnish tangos already showed local color. The song "Lumihiutaleita", we read in music textbooks and even in tourist guides of Finland, introduced "the typical Nordic imagery into the tango", whatever a typical Nordic imagery is (certainly, in the specific case, it is about replacing the passionate sexually-charged themes of the Argentinian tango with the topics of longing and nostalgia). And since then, the Finnish tango has struggled (rather successfully, by now) to be granted the label of authenticity. But it is a rather rare case: Italian bossa nova, Japanese ye-ye beat, and Spanish hard-rock (to mention three peculiar cases) do not seem to enjoy a similar privilege, despite the fact that a composer like Ennio Morricone wrote sheer international bossa nova standards (take the theme for the *Love Triangle* movie, to mention one), a band like Pizzicato Five have been among the most clever beat acts of the 1990s, and a band like Heroes del Silencio produced top-quality hard-rock.

If we abandon popular music for a second, a rather poignant case in classical music is the quest for authenticity related to the rise of national schools in the 19th century.

History. Another rather obvious aspect is the opposition past/present, with its diverse nuances. The most typical of these refers to the time when a given type actually originated. That principle also includes the opposition original/remake (or re-arrangement, transcription, cover, etc., depending on the medium and the stylistic context). However, the opposition past/present is not the sole one to be displayed on temporal units. Temporal dualisms can be expressed also (in fact, more properly) in terms of before/after, and there can easily be occurrences where the two elements of comparison are both located in the past. For a number of reasons there are, in music history, times and periods that are considered more authentic than others, as their intrinsic socio-aesthetic characteristics are claimed to have displayed more of the above-mentioned ethical connotations than the ones to which they are compared (the dualism baroque-rococo being a typical instance, and of course the so-called authenticity movement makes sense exactly in this respect).

I was once browsing the web-pages of a band called Rufus Stone: "The music is authentic 60s soul with selections [...] from the soul greats such as Otis Redding, James Brown, Wilson Pickett, Joe Tex and many others". Wow, I thought, it really sounds like a good band. Unfortunately Rufus Stone is a *cover* band, i.e., a group that performs only songs written by other authors. It is significant that many of the musicians using the label "authenticity" as applied to their own music are in fact those who have an inauthentic approach by definition. But yes, authentic soul arguably comes from the 1960s, the golden age of such labels as Motown, Atlantic or Stax. Or does it? Did Stevie Wonder fool all of us, by producing his (by far) best output in the 1970s? "I just called to say I love you", for many his (artistic) death certificate, came only in 1983: all over the previous decade, he delighted us with milestone albums like *Songs in the key of life*. How could he be so late?

Authentic punk music should be dated no earlier and no later than 1976–1977, yet The Clash delighted us with *London Calling* (an acknowledged masterpiece of the genre) only in 1979. In the same year, one of the most important progressive albums of all times, Pink Floyd's *The Wall*, was released, several years after prog had supposedly terminated its golden age.

The "before and after" principle includes also the opposition original/cover, as emerging when one compares—say—Bob Dylan's "All along the watchtower" with the several cover-versions issued later by musicians of the likes of Jimi Hendrix, XTC and U2. But then again somebody was cunning enough to make us believe that their own versions of previously-released songs were *the* authentic versions. Is it not so when we listen to the desperately passionate soul of Joe Cocker's performance of "With a little help from my friends", and we compare it to the reassuring baritone of Ringo Starr's voice and the clean perfection of George Martin's production in the Beatles' original? Or is it not so when we watch the late Jeff Buckley, alone with his guitar, crying out "Hallelujah", and—not before thinking about how young and tragically he died—we compare it with the relaxed gospel of Leonard Cohen's original?

But there are more extreme cases than these. Joe Cocker's cover of "With a little help from my friends" may be considered the "ultimate" version of the song, but nobody really goes as far as to ignore that this is originally a Lennon-McCartney work. There are instead songs "performed by" that become straightaway "written by" in common perception. People, for instance, tend to think that The Beatles actually *wrote* "Twist and Shout", which is in fact a Phil Medley and Bert Berns composition, or that "Everybody Needs Somebody to Love" (written in 1964 by Jerry Wexler, Bert Berns and Solomon Burke, and originally performed by the latter) was explicitly written for John Landis' movie *The Blues Brothers*, released 16 years later. And so on.

A consequence of such a shift is that the particular performance that replaces the original becomes also the point of reference for *further* performances. That is, it kind of becomes the original itself. Performers covering or re-arranging that particular song will have in mind that version of the song as the model, and forget how it all started.

The famous ballad "Every Time You Go Away" was written by Daryl Hall, and released by Hall and Oates in their 1980 album *Voices*. However, the song's best-known version was released by Paul Young in 1985, on the album *The Secret of Association*, and as single (something Hall and Oates did not consider doing in 1980), hitting the number one spot on the Billboard charts.

Now, among the many differences between the two versions, there are two that I wish to underline, as they represent the best example of how Paul Young's version became by far the "ultimate performance" of the song. First, the refrain displays a very typical harmonic progression: I, vi, ii, V. In the original version, the bass inaugurates the refrain by playing the dominant, not the fundamental (as it would be more obvious), of the I chord (thus proceeding like this: V, VI, II, V) The effect is less "cathartic" than the ordinary solution, but coherent with the lyrics, that offer no solution/happy ending to the strophes: the woman of the song is still leaving her partner (the narrator), symbolically "taking another piece" of the protagonist (that is, breaking his heart a little more). It is thus logical that the bass does not give "rest" to the song, but rather keeps the tension on. In Young's version, the bass "goes mainstream" and effortlessly rests on the fundamental.

Secondly, the vocal performance: the refrain's phrasing of Hall & Oates' original is pretty much in legato ("Everytime you go away—*pause*—you take a piece of me —*short pause*—with you"). The word "away" occurs on the second accent of the refrain's second bar. In Young's version, the refrain presents a significant change. Now it is "Everytime you go—*short pause*—away—*pause*—you take a piece of me—*short pause*—with you". That is: the word "away" is sang after an additional pause that was not there in the original song, on the third accent.

So, what happens after Young's successful release? It happens that all subsequent adaptations of the song (notable ones include Barry Manilow; Clay Aiken; Al Jarreau and George Benson; Gloria Gaynor; even Richard Clayderman) take Young's arrangement as a model, and completely ignore Hall and Oates's original. Whatever new variation is introduced, it is based on Young's hit. The proof is in those two features (and others I shall not mention here): the refrain's bass line (with the fundamental on the first chord) and the extra pause between "go" and "away" are always there, whatever re-elaboration of the song we listen to. The liner notes of Al Jarreau and George Benson's album *Givin' It Up*, recalling how much they liked "Paul Young's song", leave no doubt that most people by now do not even know that a previous, original version of the song actually existed.

Finally, let us not forget the most typical attitude (a cliché, in fact) of the whole category: the fans' appreciation for the "early work" of their idols, i.e., the time when they were artistically pure, and free to express their talent without the constrictions of the music business (see also point 4 of this list). Is there a single, and I mean a single, Genesis fan who is willing to say that after Peter Gabriel quit they actually became a better band? How about the Rolling Stones, Brian Jones-, or even Mick Taylor-era, as compared to the line-up of nowadays? And Oasis' first two albums (for some even only the first one, *Definitely Maybe*, then that was it), as compared to the sadly repetitive recent output (and break-up, commented by a fan with an unmistakable: "it's ok: their late albums were s*** anyway")?

Style. Diverse oppositions emerge also in the realm of styles and genres, which we might define as a part of the level that Tarasti names "the inner properties" of an artwork (Tarasti 2000: 125). Given tokens of such types are usually claimed to be more authentic than others, probably because they seem to be provided with those intrinsic characteristics that I mention in points 4 and 5 of the present list. That applies also to the macro-categories (like rock, blues or reggae). Even when not tied to genres, stylistic choices are still passable to be labeled as authentic or inauthentic, as in the case of the opposition acoustic/electronic, period/modern instruments, etc.

Given musical genres (like garage, punk, rap and others) are usually claimed to be more authentic than others. Will there ever be a chance for synth-pop to be considered authentic? How lovely was Hugh Grant's parody in the *Music and Lyrics* movie, with that celebration of 1980s inauthenticity embodied in the song "Pop Goes My Heart"?

Still in the category of stylistic choices, I shall also put the opposition simple/elaborated, as applied to arrangements. In folk music, a basic guitar-and-voice, "first-take", "plug-and-play" set-up are often considered more authentic than complex orchestra-based multi-tracks arrangements. In that sense, the white middle-class pompous versions of folks and ragtime provided by Tin Pan Alley in the early 20th century (see the crooning style or most Irving Berlin's output) is often referred to as the inauthenticization of such styles.

Economy. A further level to be considered is the socio-economic one, i.e., the impact of music on people in terms of social phenomenon and (when it is the case) financial business. Here, we find more dualisms evaluated in the form of ethical statements, and—once again—the issue of authenticity emerges. One case is the opposition between alternative/avant-garde/marginal/cult and the so-called "mainstream", that is, music produced and distributed on a very large scale. Predictably, the former group is often claimed to be more authentic, basically because it did not have to deal (and clash) with the cynical rules of show business. Such an attitude can be extended also to the opposition indie/major, referred to small, independent recording companies or studios, as opposed to big, multinational ones (like Sony, Columbia, EMI[2]...). And, of course, it is also referred to certain marketing strategies that bring the likes of Luciano Pavarotti to "mix" with mainstream-pop musicians.

The Commitments, arguably Alan Parker's best effort in his various attempts to represent popular music in movies, was all centered on a "socio-economic" quest for musical authenticity. As many will remember, the film protagonist is a young Irish wannabe manager who assembles a group of working class musicians and leads them to short-lived success in Dublin's clubs. He is very specific on wanting authentic music for an authentic social class, so (much to the film's comical effect) he proposes the band (an Irish group of unmistakable Caucasian ethicity) to play Afro-American soul and RnB, a genre that could not be further from their ethnic-geographic-historical background. In what is probably the most hilarious sequence of the movie, the manager attempts a cultural brainwashing of the band by playing a video of James Brown and inviting them to repeat after him "Say it loud: I'm black and I'm proud" (with the even funnier effect of pronouncing the slogan with a heavy Irish accent: "Say it lo-od, I'm black and I'm pro-od").

Performance. The last category refers to the level of "action" of the artwork (with which Tarasti deals in a whole section, at pages 125–128 of his book). The given way to perform a work can be perceived as more authentic than another (and, here, criteria are legion: spontaneity, faithfulness to the composer's intention, a certain tendency not to "over-play", etc.). In popular music, this is also the case of live performances, as opposed to studio recording, or to musicians that are perceived as not really "playing" their music, like DJs.

The opposition studio versus live performance (which is, 99 % of the cases, an Oa vs. Ob situation, and not vice versa) is a good opportunity for discussing even more the question (already mentioned in the "history" point of this list) of "Ob" becoming an acknowledged "Oa", in most people's perception. The great Italian singer-songwriter Sergio Caputo once said that "songs are like boxers: they need to be repeatedly punched in front of an audience before they get the final shape". It is a fair comparison (which surely includes also the possibility that—at the end of its career—this "boxer", evidently a strong one, sounds exactly as it did at its very beginning, before it ever stepped in a "ring"): performances are hardly "final". It is in the nature itself of "performing" arts to be "in progress". Theatre, dance, music and the forms of art in-between these, differ from cinema or visual arts[3] precisely for their inherent dynamicity. However, the implication in Caputo's statement (and

[2]It must be mentioned, at this point, that the emergence of Internet-distributed art (particularly on such platforms as MySpace, YouTube, P2P file-sharing software) is causing a true revolution in this context. Music is certainly at an advanced stage of this process (mostly because of the nature of the medium itself).

[3]I am of course talking in general: I am perfectly aware that the line cannot be drawn so radically, especially when we think of gray areas like interactive art, living sculptures, and so forth. Then again, there are scholars ready to swear that any art involving performances is just another variation on the idea of theatre. Be that as it may, the point here is that there are artworks whose existence relies on performances, and others that are fixed in time/space/action.

in this section too) goes a bit further, that is, Does this dynamicity go as far as to actually create more *identities* of a performance (that is, identities that are in fact capable of putting into question the supposedly original one)?

Unless we intend to argue that demos and similar drafts are already one form of identity (they may be or not, it is not a relevant question now), we may start by considering the performance committed on record as the primary type of identity assumed by that text—the actual Oa. We know that this is not always the case, but we may safely say that, in this historical moment and when we discuss a prototypical form of popular music, this appears to be more a rule than an exception. A song, the primary text of popular music, tends to be "introduced" to an audience by its studio recorded version.

After this stage, a few things may happen, among which we could mention:

1. That song is considered particularly radio-sexy, and has potentials to become "a single", or—now that *singles* as such have lost most of their commercial impact —the song with most airplay and promotion. In this condition, it is not unlikely that the song will be re-edited (shortened, most of the times, to fit the golden 3–4 min format: Hall and Oates' number one song "Say it isn't so" suffered a good 2 min diet before becoming a single[4]), or even re-arranged (generally "enriched", in order to acquire more colors and "stand-out" qualities: a comparison between the single-edit of U2's "Elevation" and the album version will perfectly clarify the point). In this restyling process, the chances that the "ultimate" version of that song will be the one originally committed on record are very low. Later re-issues of the track, indeed (particularly "Best of" collections), tend to convey the message that the single-edit is *the* performance of that song.
2. Not so often, but still in a few cases, the song does not particularly "work" in its initial release or version, and—instead of edited or re-arranged—is more likely re-made (or re-arranged *very* heavily[5]). Historical examples include the fateful idea (by Simon and Garfunkel's producer Tom Wilson, who acted without even informing the duo) to add poignant electric instrumentation to *The sound of silence*, re-releasing it on the eponymous album, after the unsuccessful inclusion in the previous *Wednesday Morning 3 a.m.*, and providing the breakthrough for what will soon become the most successful duo in music history. Also, the Beatles' decision of recording another "Revolution", faster and electrified, for the single release, after they had recorded what came to be known as "Revolution #1" (a laid-back, mostly acoustic song), belongs to this category.
3. More often than any of the previous examples, a song gets a "live" rendition. In concert, in TV appearances, in unplugged shows… wherever. If that song becomes part of the live repertoire of a performer (and especially when it is

[4]Particularly (I suspect) because in the original version the refrain starts after no less than 2 min from the beginning of the song. Commercial suicide.

[5]Perhaps I should here specify that to re-arrange a song means to operate (with over-dubbings) on the existing recording, while to re-make it means to make a new recording from scratch.

re-proposed in more than one tour), then its destiny is the one of Sergio Caputo's metaphor: the song is officially "a boxer".

When a performer/band decides to play a given song in concert, they have to face a number of strategic problems, the most important of which is the degree and the quality of "correspondence" (i.e., as we shall see later: engagement or disengagement) between the recorded and the concert version. How *faithful* will the live version be to the studio version? Answering this question basically means taking into account all the Greimasian modalities (must/can/will/know). For instance the performers will have to compare the personnel employed for the recording to the personnel playing in the concert: *can* they do everything that was committed on record (which was likely benefiting from multi-tracking and over-dubbing procedures)? Then, they have to consider what kind of rendition of the song they intend to offer to the audience: do they *want to* be faithful to the recording, or do they prefer, say, an acoustic version, a longer solo, a different intro? Then, they have to consider what role, in their career, does that song play: do they *have to* play it in a certain manner because the audience expects it that way, is it an unavoidable moment in their concerts…? And so forth.

To this, one must add the specific conditions (including rituals) of a live performance. There are codes and conventions related to performing a song live, and these have to do with

(a) the specific context: big or small venue, home or abroad, particular significance—if any—of the venue (do these performers have a special connection with this town[6]? Is it their hometown? Is it a place named in a song? Will they have a guest-star here?), etc.;
(b) the conventional performing techniques of a live concert: in the case of blues-derived genres (rock, pop, RnB…) we have for instance an outstanding increase of *Yeah*'s, *C'mon*'s, *Alright*'s and interjections of various nature in the singing parts;
(c) the image of the performers: in general and in *that* moment (are they going through a new stage of their career? Are they going back to their roots? Is there a new member in the band? Did somebody die? Did they tell something awkward to the press? Did they just top the chart for the first time? Did they split? Did they re-form? The list is really long);
(d) the performers' relation with the fans: are they giving their best in concerts? Are they talkative in-between songs? Do they have the inclination of letting the crowd sing their songs' refrain? Do they invite adoring fans on the stage? Etc.)

The rich combination of all these features leaves very few doubts that the live-version of a song is virtually a *new* text. And even the word "virtually" can be removed in those cases when (for various reasons) the song is indeed given a new

[6]That is: are Lynyrd Skynyrd playing "Sweet Home Alabama" in Alabama or somewhere else? Is (was) Frank Sinatra singing "New York New York" in New York or somewhere else? Etc.

significance (e.g., see the case of "Every Breath You Take" on the Live 8 event,[7] but also heavily re-arranged songs that become new texts: a good example is Eric Clapton's slower, bluesy and relaxed acoustic version of "Layla", as compared with the original version released by Derek and the Dominos[8]).

Not infrequently, finally, this new text may become a *dominant* text, that is, dominant over the studio-recorded version, or any other version, and left to posterity as the "ultimate" rendition of that song. It is then common, in those cases, that even those little conventions of live performances crystallize and end up becoming basic (rather than ornamental) ingredients of the text.[9]

I tested this point by comparing the live repertoires of Paul McCartney since the breakup of the Beatles to nowadays (which covers 45 years and 20 tours, of which one is ongoing at the moment I write and 15 are "World" tours, i.e., outside the UK). On these occasions, more than one song was treated to this text-reconfiguration I described, and that is more unexpected that it may sound,

---

[7]The Live 8 benefit mega-concert was organized in 2005 by Bob Geldof in order to raise consciousness of the *Make Poverty History* campaign among those world politicians who were then about to meet in Edinburgh for the so-called G8 summit. On that occasion, "Every Breath You Take", performed by Sting in a way all in all similar to the original Police recording issued on the *Synchronicity* album, and fairly unchanged in the lyrics (if not for those variations typical of live performances, with some additional "yeah!" and similar utterances), completely turned upside down its formerly love song contents in correspondence of the coda "Every breath you take, every move you make, every step you take, I'll be watching you". At that point, big photographs of the very G8 politicians would be displayed on the various big screens of the venue. Now, the lyrics clearly achieved the features of a protest song, in which it was the politicians' steps, moves and breaths to be carefully watched.

[8]A song like "Layla", in the mentioned two versions, is almost bound to change significance, regardless of its arrangement. The song, first released in 1970, was a secret love declaration for Pattie Boyd, then wife of Clapton's best friend George Harrison. The very awkward situation of being in love with his best friend's wife was a good reason to call the female character of the song Layla, instead of Pattie (never mind that the former musically sounds much better): in fact, the specific choice of "Layla" was motivated by a novel called *Majnun and Leylà* where the characters face a very similar situation of the triangle Harrison-Clapton-Boyd. Now, the Derek's version presents features of anger and desperation (music critic Dave Marsh praised the song by saying that hearing it was "akin to witnessing a murder, or a suicide"), certainly motivated by the situation itself (at that point, Boyd was determined not to leave her husband), and possibly fuelled by Clapton's serious drug-addiction of the time. History has it, Boyd would divorce from Harrison in 1977 and marry Clapton in 1979, but, in turn, the couple, after Clapton's repeated adulteries and even a son born in an extra-marital affair, divorced ten years later. In 1992, three years after the divorce, Clapton performed the song in the new acoustic version for the MTV Unplugged show. Emotionally-distant and divorced from the woman he so much seemed to love 22 years back (and also cleaner and more sober than in those days), Clapton was clearly in the position of delivering a much more relaxed and understated version of the song. In a way, Layla became Leylà for real, that is, a fictional character involved in a love triangle that could be just anybody, and not necessarily the result of the author's autobiographical accounts.

[9]Italian singer Vasco Rossi has created a sort of sacred ritual in the live-performances of his signature song "Alba Chiara", by constantly adding a long "Hey!" after the verse "sei chiara come un'alba", that was not present in the studio-version of the song, and that the crowd utters loud *every single time* that song is played in concert.

because McCartney belongs to that school of performers who does not particularly love to restyle his songs, and rather prefers to play them as faithfully as possible to the original.

Although I cannot check with meticulous care (I possess playlists from every tour, but not from every concert, and it is not rare for performers to add or subtract numbers during the same tour), I feel fairly confident in stating that one of the songs that McCartney has played most often during these 20 tours, is "Maybe I'm Amazed". A somehow unexpected choice, when one thinks that we are talking about the same author of "Yesterday", "Let It Be", "Hey Jude", "Band on the Run" and dozens of other evergreens. However, looking closely, a very reasonable one in the light of McCartney's personal and professional history within the *after*-Beatles scenario, where "Maybe I'm Amazed" stands as

1. a song written in 1970 and released on the very first solo album of the English singer-songwriter (therefore, in principle *available* for concerts already from the first tour, the famous University tour of impromptu gigs of 1972—where however it was *not* featured);
2. the first post-Beatles *classic* that McCartney managed to write, therefore of crucial emotional and psychological importance for regaining self-confidence after the traumatic breakup of the Fab Four[10];
3. one of his best non-fictional sets of lyrics: a very personal song about his wife Linda, at the beginning of their 29-year-long marriage, with a direct, vulnerable and sincere attitude that was more typical of his ex-songwriting partner John Lennon (while McCartney, at his lyrical best, has always been a Dickensian storyteller, creator of a multitude of completely fictional characters and events[11]). In this sense, the song had obviously a great significance in McCartney's repertoire, plus—a rather relevant detail—it proved from the very beginning to "go down well" live, embodying the epic features of the arena-rock ballad.

Be that as it may, "Maybe I'm Amazed" appears in the regular playlists of no less than 17 out of 20 McCartney's tours. It only missed the very first tour, the short unplugged tour in 1991 (possibly for the very reason of featuring crucial electric parts, like the characteristic guitar solo), and the one in 2010. It is an impressive presence, that exceeds any of the Beatles classics (which he started introducing in his sets only from 1976 onwards, after a period of conscious refusal of identification with his past), and perhaps any other songs from his vast repertoire, with the possible exception of the sole *Live and Let Die*, a must in his concerts due to the spectacular fireworks used during the instrumental parts.

[10]It should be remembered that, despite being the one who legally dissolved the Beatles partnership, McCartney was the Beatle who last surrendered to the idea of splitting the band.

[11]Dickens is famously estimated to have invented more than 13,000 characters. It would be interesting, one day, to count the various Eleonor Rigby, Desmond and Molly Jones, Sergeant Pepper, Maxwell Edison, and see how many characters were actually invented by the ex-Beatle.

Now, back to our question, is the live-version of "Maybe I'm Amazed" that sort of new text that managed to become a *dominant text*? Playing with words, does McCartney play a live-version of the studio-version, or a live-version of a previous live-version? What text does he refer to when he plays the song nowadays, or in 2002, or in 1989, or in 1979 (and therefore, possibly, *which* "Maybe I'm Amazed" prevails as *mental map* when people think of, or sing, it)?

The answer is: the live-version of the Wings Over the World Tour of 1975–76 is the "ultimate", dominant, version—the Oa of the situation (or shall we say "O alpha", since we are talking about dominance?). That is, not the original studio-version appearing in the 1970 album "McCartney", not the first live-version from the 1972 UK tour (McCartney's second tour after the Beatles' breakup), which is a clearly transitional version between the original and the one I am about to discuss.[12] What happened on the Wings Over the World tour to that song was that it became the ultimate version that McCartney, almost invariably, performs every time "Maybe I'm Amazed" is played live. All the variations, additions, liberties introduced there became "standard", and very well known to the audience as well, who *expect* those things to happen (up to the point of singing along to them) and not the ones that occurred in the *McCartney* album.

This example is significant because it is recurrent but not regular in McCartney's performance history: there are songs subject to the same phenomenon (another one is "Jet", again crystallized on the 1975–76 tour), and songs that are not (take "Every Night": deep and powerful on the 1979 tour, light and *à la* CSN&Y in 1993, and voice-and-guitar-only in 2002). Such transition of "power" from one text to another is analytically relevant because it may or may not happen. It is not an automatic process, and neither one too rare to be worthwhile of attention.

Therefore, what are the differences between the studio and that particular live version of "Maybe I'm Amazed"? Why did the new features become dominant? And, finally, did more recent renditions change the picture *at all*? Let us look at these issues in detail, by starting from a comparison between the two texts (we shall from now on call SV the studio version, and LV the live version):

1. *Length*: SV lasts 3 min and 50 s. LV lasts 5 min and 20 s.
2. *Tempo*: the main reason of this minute-and-a-half difference between SV and LV consists in the significantly slower tempo of the latter. We switch from the roughly 80 bpm of SV to the 70 circa of LV. It is a rather large gap.
3. *Coda*: the other important reason for the difference in length is the existence, in LV, of a coda that was completely absent in SV. Melodic and harmonic elements of that coda do not appear in any shape or form during SV. It is literally a new part in the song.
4. *Ending*: as customary of live songs, LV has a definite ending, with the whole band "finishing" the song. SV, on the contrary, fades out on one refrain (LV completes that refrain, and—as mentioned—launches the coda).

[12]Providing details on this transitional version, too, might make the text a bit too demanding for what, after all, is just an example to clarify a point. Briefly, it should be said that this early live-version has roughly the same length of the studio-version, but already introduces (although shorter) a coda that will characterize the "dominant" live-version of 1975–76.

5. *Beginning*: SV starts with a fade-in (a technique, incidentally, that was introduced in rock exactly by the Beatles, in their 1964 song "Eight Days a Week"). LV, as with the ending, has a clear opening, which roughly coincides with the very piano chord that can be distinctively heard from the fade-in in SV.
6. *Instrumentation*: the most relevant difference is the absence, in LV, of a second guitar that instead appears in SV. McCartney's live bands always had two guitars in their line-up: however, whenever McCartney would switch from his bass to a piano (and such is the case with "Maybe I'm Amazed"), one of the two guitarists would replace him as bass-player, leaving just one guitar for that performance. The electric organ clearly audible in SV is instead a feature that McCartney reintroduced in LV only from the 1989–90 tour, and constitutes one of the extremely few "evolutions" of LV over time.
7. *Refrains*: SV features four refrains (the last one fading-out the song), the first two of which are sung with lyrics, while the remaining two are accompanied with cries of "woah". In LV we have four *complete* refrains (no fade-out), *three* of which are sung with lyrics, and one (the third one) is filled with non-linguistic cries of different sorts.
8. *Melody*: the structure of cries and interjections apart, the melody has a significant variation in the refrains. In SV the first and the second part of the refrain ("Maybe I'm a man, maybe I'm a lonely man who's in the middle of something…" and "Maybe I'm a man, maybe you're the only woman who could ever help me…") are different, in that the second hits higher notes in correspondence of the "…could ever help me…" passage. In LV, with the exception of few performances, the first and the second part of the refrain are identical in melody.
9. *Singing style*: in SV, the voice tends to be rather clean during the strophes and bluesy hoarse during the refrains. In LV, the timber is constant across the different parts, and settles on a middle way between the two in SV.

To this, we shall obviously add all these live conventions that we mentioned. Once again, it is difficult not to note how numerous *Yeah*'s, *Ooh*'s and falsettos are added in LV. It is worthwhile noting that it is the third refrain, the only one that, in either SV or LV, is not performed with lyrics but with non-linguistic vocalizations. Well: in SV those vocalizations have a mild and disciplined role, and basically serve the purpose of reinforcing the upbeats (where the snare drum sounds). In LV we have a whole catalogue of vocal sounds and melodic excursions that are more like a "solo". As it happens, the very same structure of this vocal solo was repeated, virtually unchanged, in all McCartney's following tours, and—as someone who attended four concerts on three different tours—I can testify that the audience is perfectly familiar with these vocalizations and tends to sing along to them, as if they were always there.

And so on. All of these features, in the LV configuration, became standard, in McCartney's performances as in the audience's mental maps of the song. There is hardly a doubt that the "ultimate" version of "Maybe I'm Amazed" ("the definitive take", to put it like music critic Stewart Mason from the AllMusic portal), is the one that was perfected in 1975–76. It is very notable that, after the live album from that tour came out (*Wings over America*, in late 1976), "Maybe I'm Amazed" was the song chosen as a single (reaching the Billboard Top Ten: not an usual instance for a live-version of an already released song): a clear recognition, on McCartney's part, that this was a stand-out track. Back in 1970, none of the *McCartney* album songs had become singles.

As we said, with only three exceptions, "Maybe I'm Amazed" has become a true war-horse for McCartney, both at the time of Wings and in more recent years, when—from 1989 onwards—the ex-Beatle resumed intense and regular touring activity. In these 45 years there have been numerous line-up changes in his touring bands, either partial (one or two elements) or total, and in the end we can notice that there have been three integrally different bands (Wings in the 1970s, a backing band in the 1989–1993 period, and another backing band from 2002 to nowadays), and—if memory serves me—an overall rotation of no less than 17 musicians. To all of them, the task has always been the same: "Maybe I'm Amazed" shall be played with that tempo, that coda, that instrumentation, and so on. The "point of reference" is the 1975–76 version. Yeah's and Ooh's included! In fact, there is a rather revealing sequence in the concert-film *Back in the US*, which documents the 2002 tour, where Paul "Wix" Wickens, the keyboard player of McCartney's current backing band, is actually "conducting" with gestures and facial expressions the "traditional" set of vocalizations that McCartney delivers in the third refrain. The thing is, Wickens is not actually conducting: he is "following" what he knows will happen, and, if anything, "encouraging" it too.

To conclude, is anything actually *left* for any following performance of "Maybe I'm Amazed", or any other song that has acquired a dominant role in its live version? Of course yes. What we have discussed so far is nothing else than an elaborate way to agree with Donatas Katkus, when he provides a definition of "performance" that suggests the existence of a plan, or at least a draft (as in improvisation) behind the music-making:

> The Lithuanian word "atlikimas", as well as the Russian "ispolnenije", both stand for "execution", that is, an act that embodies a set of previously-planned conditions. The German "Aufführung" and the English "performance" too are related to the execution of a previously-planned job. All these terms show that, in embarking on the act of musicianship, something is indeed performed, that is, a certain predetermined idea is put into action, or a certain musical structure – like a song, a dance, a chant, or a symphony – is repeated. The term "to perform" has replaced the previously-used "to sing", "to play", "to make music" etc., in order to name the concrete act of musicianship and its nature. (Katkus 2006: 14[13])

Performances, indeed, feature "previously-planned conditions". And, as we have seen, not infrequently the "previous plan" becomes a standard plan and the main point of reference. Along with a planned performance, however, we also find what we might call a *hic et nunc* performance, that is, performative elements that are

[13]I am grateful to my wife and musicologist colleague Lina Navickaitė-Martinelli for translating Katkus' text for me.

extemporaneous, suggested by specific conditions occurring in real time and by all means *not* planned. The repeatedly-mentioned vocalizations, in a normal situation (the third refrain in "Maybe I'm Amazed" is indeed more an exception), are ritual *in principle*, but in practice they may happen at any point of a song, in different disguises, and hardly reveal a pre-existing intention. The same goes for plenty of other elements: more typically lead vocals, solos, drum-fills and embellishments; less typically rhythm patterns, backing vocals and other things in the "background", but here, too, there is no rigid rule and things may happen in different combinations. These occurrences provide the performance with a quality of uniqueness, an unrepeatable collocation in *that* space and at *that* time.

Which, at the end of the day, is and remains the basic, unavoidable difference between forms of art that imply performances and forms that do not.

Further dualisms that should be included here in principle (imperfect/perfect, dirty/clean…) will be discussed further on, in the paragraphs on "Error Aesthetics".

Rhetoric. A last form of categorization, transversal to the ones so far discussed, puts the relationship authenticity-inauthenticity on a more syntactic level, particularly on the rhetorical dimension. In other words, the attempt is to describe the *appearance* of this relationship, its formal strategies. In this case, and just to mention a few examples out of many, we may discuss at least the following four tropes: "paraphrase", "metaphor", "synecdoche", and "citation". I have devoted a specific treatise on this subject in an essay on Beatlesesque repertoires (Martinelli 2010b: 159–160), therefore I shall draw my examples from there:

1. A *paraphrase* occurs when Ob is clearly imitating one specific Oa, generally for purposes of parody, homage, philological reconstruction, or other. In Monty Python's irresistible parody of The Beatles, *The Rutles*, we find an enormous number of songs conceived as paraphrases of existing Beatles' ones ("Ouch!" for "Help!", "Get Up and Go" for "Get Back", "Love Life" for "All You Need Is Love", etc.);
2. A *metaphor* occurs when Ob ideally (metaphorically) reminds us of Oa, without having any particular, specific resemblance. The purposes are similar to point 1, except that, obviously, the philological approach is replaced by a different, consciously vaguer, more aesthetic strategy. "Pineapple Head", by Crowded House, is clearly similar to "Norwegian Wood", but the question here is mainly a matter of 'atmosphere' (the 6/8 beat, the acoustic guitar riff, and so on) rather than a specific feature (unlike a paraphrase, one cannot speak of a metaphoric song in terms of rip-off, but rather of inspiration). For this reason, a metaphoric Ob does not need to have a specific "relation" to Oa (parody or homage), and can easily stand on its own as an original artwork;
3. A *synecdoche*, here meant in the "part for the whole" form, occurs when Ob refers to an entire class of authentic objects (school, genre, form, etc.). "Whatever", by Oasis, is a song where one can hear echoes of "Hello Goodbye", "I Am the Walrus", "For No One", and some other, without necessarily enabling the listener to keep these influences separated and recognize these songs individually at any point in the Oasis' track. A synecdoche is usually the most

common kind of Ob, and possibly embodies the idea itself of inauthenticity in a quintessential manner;
4. A *citation* occurs in two different forms: (a) when (in a synecdoche context) Ob presents a specific part/s that call to mind a specific part/s in Oa (the middle-eight in "Country House" by Blur is clearly an homage to the middle-eight in "Doctor Robert"); and (b) when there is a specific part/s of Ob, reminding us of Oa in general, which is now "quoted" (as a style) in a context that otherwise is not necessarily pertinent to it. Back to our set of Beatlesesque examples, the band XTC were quite specialized in this respect, applying isolated catchy-Beatlesesque passages in songs that otherwise have a more idiosyncratic character (as is typical of this great, but underrated, band): appropriate examples of this kind could be "You're The Wish You Are I Had" (with its Beatlesesque refrain) or "Great Fire" (with its Beatlesesque middle-eight).

3.2.2 Engagement–Disengagement

Summarizing what we have so far we can say that in order to apply the notion of authenticity within a discourse we need at least two signs (A and B) in reciprocal interaction, and this interaction is possible via a number of both fixed elements and variables. In Eco's model the variable is t, the "historical circumstance", while subjects and objects are to be considered constants.

Other important scholars have discussed the notion of authenticity. Among these, the Lithuanian semiotician Algirdas Greimas whose theories are in fact the actual basis of Tarasti's application of the notion to music:

> Applying Greimas again, one can speak of authenticity of time, place, and subject (or "actor"). Centrifugal and centripetal forces operate these three dimensions. Greimas calls the centrifugal force, which makes a text move in the inner or outer sense, débrayage (disengagement), and the centripetal force embrayage (engagement). Theoretically, authenticity is at its greatest when complete engagement prevails in all three dimensions: the temporal now (nunc), the spatial here (hic), and the actorial I (ego). Such a state has always been considered the ideal; it is the utopia of philosophers. (Tarasti 2000: 118)

Now, not only such a model justifies the criterion of classification proposed in my list (giving, for instance, full credit to the spatial and temporal dimensions of the previous paragraph's points 1 and 2). It also appears slightly more exhaustive than the one proposed by Eco, for at least two reasons: first, it is explicitly aware that a notion like "historical circumstance" is incomplete (or at least, it leaves no ground for misunderstanding in that respect); and second, it is flexible, in that it allows more articulated interpretations of the phenomenon of authenticity in the arts.

To start with, the notions of ego, hic and nunc are not as stable as one could expect. I have already mentioned the case of musicians re-performing their own material in live contexts (or remix albums, or else), but the issue is even more complex. What is an authentic piece of music? What is the entity one should take as a point of reference? For quite a while, in classical music, the score was that entity most of the times. However, nowadays, the absolute majority of musicians (experimental, popular, jazz, electronic, etc.) do not start their compositional process by writing a score (in fact, some of them declare they are not even able to write music). They instead use a piece of paper and write down lyrics or other constitutive elements of the song; they might use a demo-tape, a computer file, or any other draft (including incomplete representations of the work: many songwriters use a sheet of paper to write lyrics, and rely on memory for the melody).

More to the point, the very 'thing' that ends up in the hands (and ears) of the receiver (the listener, the fan, the music-magazine journalist, etc.) is, or may be, a product that has very little to do with these approximate drafts. The history of popular music is full of cases where the song released on the record was significantly different from the "original". The original Doors song "Light My Fire" was actually a folk*ish* ballad without that very famous (and impressive) electric organ riff (only later provided by Ray Manzarek), with a different melody in the verses, and far shorter in length than the one we all know. Supertramp's "Breakfast in America", in the original demo by Roger Hodgson, had a totally different rhythm and no refrain (the latter eventually becoming the most characteristic passage of the song). The original lyrics for the Beatles song "Yesterday" (originally called "Scrambled Eggs"), went like this: "Scrambled eggs, oh my baby how I love your legs…", and the original demo was recorded with voice and piano. Are these the points of reference for a scholar who analyses "Light my fire", "Yesterday", or "Breakfast in America"? When we analytically compare Gounod's "Ave Maria" with the first "Prelude" from Bach's *Well-Tempered Clavier* (as compositions, not as performances), we normally do it on the basis of their respective scores, relying also on historical notions (like, in this case, the fact that Gounod composed the piece *on top* of Bach's prelude). The scores are what we take as "the truth". However, when it comes to comparing Ray Charles's cover of "Yesterday" with the original song, composed by Paul McCartney, our point of reference cannot possibly be "Scrambled Eggs"; nor can we simply stick to a score that McCartney certainly did not write, and was only eventually published in its entirety. In fact, what entitles us to speak of Ray Charles's cover as a 'cover', is its primary performance, i.e., that very recording called "Yesterday", issued in 1965 within an album titled *Help!* credited to a band called The Beatles, produced by a gentleman called George Martin, featuring lyrics such as, "Yesterday, all my troubles seemed so far away…" and recorded with one voice (doubled, thus filling two tracks), one acoustic guitar tuned one tone lower (in order to play an F major like it was a G major, and so on), and a string quartet. None of these instruments were on the demo, no piano was in the final recording, and no mention of scrambled eggs appeared within the lyrics. Yet, this is what 99.9 % of the people one might ask to, would indicate as the authentic "Yesterday".

What does this whole picture suggest? Certainly, most of the reflections one may propose were highly anticipated in the very famous essay by Walter Benjamin, "The Work of Art in the Age of Mechanical Reproduction". In a few words, the advent of devices that allow a totally faithful and potentially limitless reproduction of an original artwork makes authenticity a rather relative concept. It is not the purpose of this essay to deepen this issue, but—within the framework of this "numanistic" critical deconstruction of the concept of "authenticity"—we may still deliver a few open questions, hoping to prompt the reader into further reflection: Is irreproducibility and uniqueness the very essential feature of authenticity? What really counts as authentic, the (chronologically) original intention of the musician, or what s/he actually accepted as being worthwhile to present to an audience, after a clear process of refinement? How much "refinement" is allowed (conceptually and chronologically) before we trespass on the border of inauthenticity? And finally, if the issue is (and it is) of an ethical nature, what are the ethically relevant features for an artwork to be claimed as authentic? Is the ego/hic/nunc issue all about answering questions like: Who did it first? When did it happen first? How is it perceived? Etc. Or is there more, and maybe concepts like mediation, shaping and multi-articulation should be fully taken into account?

There is more in the Greimasian model on authenticity. The notions of engagement and disengagement open interesting analytical opportunities. Indeed, the connection between authenticity/inauthenticity and engagement/disengagement is definitely a dynamic one. And all combinations seem to be possible.

As we are moving into a Greimasian area, it makes sense to structure the four dimensions into a classical semiotic square, as in Fig. 3.2. The square creates logical relations, which in this case are of a pragmatic type, in that they link the semantic features of the four dimensions with the syntactic ones. As these relations are analogical, I took the liberty of re-arranging the Greimasian square into a Cartesian plane that possibly (at least in a visual sense) gives more credit to the unstable space that each relation occupies (see Fig. 3.3).

Let us now see, with the help of Fig. 3.3, the four emerging relations in detail. Of course, engagement-authenticity, and disengagement-inauthenticity are the most obvious combinations. The former pair includes fully acknowledged original artworks (where it is possible to point to them with no doubt whatsoever), compared to which the object Ob (the one that raises the case) does its best to maintain the

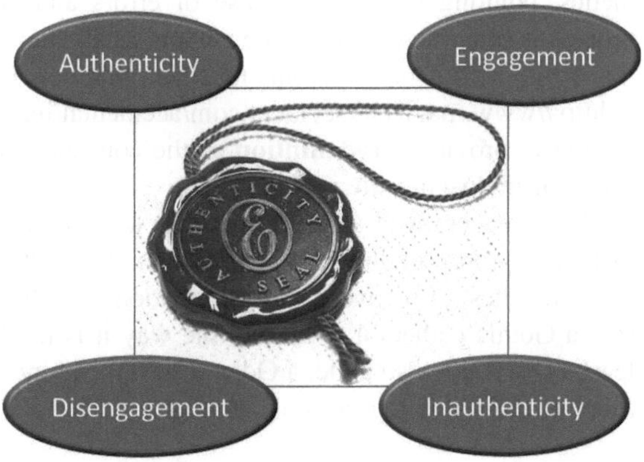

Fig. 3.2 The dimensions of authenticity/inauthenticity and engagement/disengagement, as framed in a Greimasian square

Fig. 3.3 Semantic relations between the four dimensions

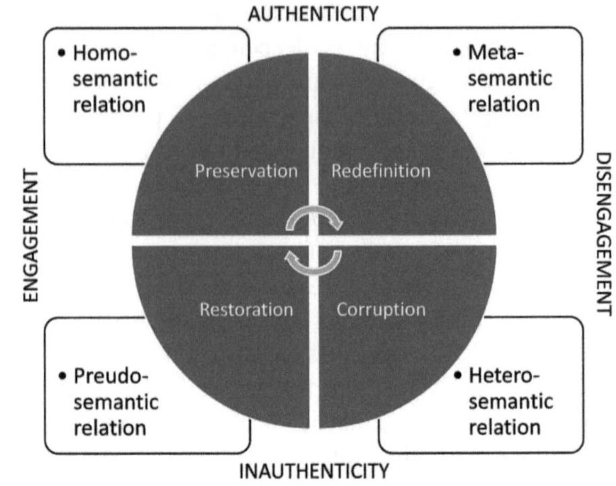

authenticity of Oa. I call this relation **preservation**, qualifying it as **homo-semantic**. In the disengagement-inauthenticity, I obviously refer to the quintessential forgeries. The relation is clearly **hetero-semantic**, and its pragmatics is a form of **corruption**.

Apparently, however, this is not the end of the story. For instance, how do we qualify the vast area of parodies, pastiches, tributes, etc.? They are of course (perceived, and sometimes meant, as) cases of inauthenticity, in that they present the basic schemes of Eco's model (Ob, rather than Oa; t2, in its already-defined more complex articulation, rather than t1; and, most of the times, a subject B rather than a subject A). However, do such cases always reflect a centrifugal tension, and thus a form of disengagement? It really does not seem so. In fact, in most cases, it looks like the exact opposite, with a force that is not only centripetal, but in fact philological as well. A combination of this type is **pseudo-semantic**, and its pragmatics is a form of **restoration**. Finally, and, however weird it may seem, it is even possible to construct a form of authenticity on the basis of a disengaging force. In the next paragraphs, I will have the chance to discuss the notion of error aesthetics, pointing out how the use of errors and accidents in the compositional process is often described by the authors as an attempt to "provide the humanity in this sterile world" (as electronic musician Matthew Herbert declares in an interview —http://www.magicandaccident.com/accidental/index.htm?2). Cases like this obviously provide a **redefinition** of the concept of authenticity, and are therefore forms of **metasemantic** relation.

Those readers who are fairly familiar with architecture may have not missed the fact that my proposed four categories are rather faithfully overlapping with the basic four strategies for recovery and valorization of architectural heritage. If we think of, say, a Gothic cathedral (Oa), and the way it is used in time (therefore creating at least a Ob, if not also a Oc, a Od and so on), we indeed end up dealing with one or more of the following possible strategies:

1. We may try to preserve, as much as possible, the integrity of the original cathedral (Oa), i.e., renouncing intervention, but also trying to keep the cathedral alive, by for instance exploiting it very little;
2. On the contrary, we might corrupt the building by adding, time after time, elements belonging to the contemporary architectural styles (no need to remind us how popular such a practice was in Baroque times, or currently is in big cities like Paris);
3. We might also opt to restore the building, i.e., to accept that a total preservation is unachievable, thus the next best thing is an intervention that tries to be as respectful as possible to the original, not avoiding, though, the use of modern techniques and technologies (e.g., frescos may be faithfully re-painted, but with more enduring colors that were not available before);
4. Finally, through a rather fatalistic action, we might accept the idea that time-passing is part of life, and that several historical events (also catastrophic ones) just happen, and therefore the most authentic way to approach that cathedral is simply following its development, whatever shape this latter takes. If, for instance, the cathedral is bombed during the war, and partially destroyed, it might simply be redefined, and become, say, a memorial monument. The building acquires thus a meta-signification that points at Oa, while being now something else.

Navickaitė-Martinelli (2014: 129–133) provided a very interesting application of this model in the field of art music performance, which is particularly relevant to any kind of study of authenticity in music (the act of performing itself having to deal, from the very start, with a potential case of inauthenticity). In Navickaitė's opinion, the case of preservation is well exemplified by the authenticity movement (the so-called historically-informed performances); the case of restoration applies to those instances where an objective approach to the score is made (Toscanini being one good representative of the category), even though with modern instruments. The case of corruption occurs in several forms of modernization and/or romanticization of the classical repertoire; and, finally, the redefinition is called for when, for instance, we witness a radical change in the medium itself (computers, player pianos, etc.) so that an actual new form of authenticity is created.

Besides Dr. Navickaitė's reflections, we can witness similar forms of relation also in the popular music environment. Engagement-authenticity and disengagement-inauthenticity, as I said, are the most obvious cases, so I do not think we need a thorough analysis of those. It shall be enough to think of the die-hard (at times slightly pathetic and anyway never fully-achieved[14]) defense of authenticity of some old bluesman or country-folk singer, and—on the opposite pole—of the most kitsch remake or cover albums. Looking closely, also the combination inauthenticity-engagement is not necessarily unlikely, especially when we think of the vast area of parodies, tributes, quotations and so forth. It is hard to assess those "inauthentic" instances as pure forms of disengagement. The truth is that all evidence leads to the

[14]As Coco Chanel once said, "there is nothing that makes you look old like trying to look young".

opposite conclusion. When we consider those many cases (again, post-modernly fashionable) of homage/revival of past tradition of electro-acoustic music exercised by modern bands and performers, there is no doubt that their aesthetic force is openly centripetal.

Such is the force that brings bands like the French group Air, the English Stereolab or the Finnish Nu Science to:

1. Use instrumentation that is certainly electronic, but mostly analogical, with such vintage objects as those Rhodes or Roland keyboard models that were very hyped in the 1970s and 1980s;
2. Adopt, as stylistic and occasionally compositional resources, technical-musical material of those same ages, including the sound-card of old Commodore computers;
3. Constructing a public image (live concerts, album covers, etc.) of a *retro-futuristic* band[15];
4. Proceed, in strictly musical terms, in a chronological direction that is clearly recognizable, calling to mind specific characters of a specific period (e.g., Pink Floyd for Air, the Krautrock genre for Stereolab, and Giorgio Moroder for Nu Science[16]).

Even in cases like musical parody, certainly not the heavily sarcastic type à la Frank Zappa, but rather the devoted and philological type of the Dukes of Stratosphear and the likes, one can witness the same "centripetal" characteristics. A careful listening to *25 o'clock* and *Psonic Psunspot*, the two albums released by the Dukes, will immediately clarify the point.[17]

However, the most interesting element within a specific study on authenticity, and in more general theoretical terms, is the most unexpected combination of the Greimasian model, that is, the relationship between disengagement and authenticity. I shall discuss this combination through what I consider its most significant manifestation in popular music (as in other arts, as a matter of fact): the so-called "Error Aesthetics".

---

[15]I use this expression to designate that particular aesthetic choice that pursues an idea of the future the way it was perceived in the past. From style exercises in *Metropolis* style, up to the re-use of stills or sequences from the old *Space Invaders* videogame, examples are legion.

[16]It must be underlined that the names I just mentioned do not necessarily pay homage to in a pinpointed, explicit way, as we see so often in nowadays pop. What I am talking about is a more general tribute, mostly expressed in terms of mere "mood", and that is once again identifiable as a synecdoche. For instance, the strong melodic and harmonic inclination of many of these bands' songs, which is certainly in opposition to the general trend of modern electronic music (the latter instead targeting a quasi-elimination of such elements, or at least a reduction of their importance), is clearly a form of engagement towards the whole electronic tradition of 30–40 years ago, and the references may be more than one at the same time.

[17]Then, after having listened to them for research purposes, I suggest simply enjoying them, as they are nothing less than two pop gems.

3.2.3 *The Notion of "Error Aesthetics"*

One of the many reasons why computers replaced typing machines in writing is doubtlessly because the former are more accurate. A text written on an old Remington is a typographically imprecise text, with several smudges in correspondence with many, if not all characters, and the impossibility of using the "delete" function in case of typods typos. Moreover, there is a wide margin of randomness in the 'capacity' of a typing machine to produce errors. A medium-quality Remington may possibly offer an acceptable perception of clarity and cleanness, but one could definitely use a machine in worse conditions (people of my, or older, generation know *exactly* what I am talking about), up to compromising the intelligibility of the whole text (which is from the start less intelligible than a comforting Times New Roman on a computer).

Having said that, however, I think we all agree that a Remington-written text possesses quite a charm. The simple fact that there are several websites where one can actually download these fonts, adapted for computer use, in their original, inaccurate and approximate form, confirms their not irrelevant aesthetic appeal. On the one hand, one can explain the phenomenon with the same old seduction of past and tradition (which, as the readers may have by now guessed, the trickiest part of the story, in our numanistic analysis of "authenticity"). This very seduction induces internet users to download—say—fonts reminiscent of sci-fi movies (remember the "Blade Runner" bending and striped inscription?), or others that are reminiscent of medieval manuscripts (as we basically see on the signs of every second British or German pub). But in the case of the Remington there is something more. Not only do these fonts try to recover the stylistic features of a given past technology: the Remington revival is also an attempt to recover the *limits* of that technology, its imprecision. Randomness, subjectivity and unpredictability are the really charming features in this case. A "Virtual Remington" font I once downloaded for my computer was designed to reproduce the characteristic smudges on a few letters; while another one, called "Old Remington" basically had some kind of defect on each letter of the alphabet, including the fact that words were never written "straight", but there would always be one or two letters slightly out of place (again, if you have used an old typing machine, you are familiar with this phenomenon). In other words, not only is the error itself attractive, but also—to a certain extent—its randomness.

To return to the main topic of this case study, in popular music a rather common strategy of "good old days" charm is to intentionally add vinyl noises in songs recorded with a technology that does not produce them (I plead guilty here, as in 2009, together with a collective of international musicians, I released a CD with my songs, in which we intentionally wanted to give an "A side" and "B side" effect by adding the noises of a vinyl player at the beginning, in the middle and at the end of the album).

In the last couple of decades, especially in visual arts, but also in the musical ones, a trend known as "Error Aesthetics" (or "Aesthetics of Failure") has been developing. It is hard to say whether this is a fully self-conscious movement, like—for instance—surrealism or futurism, or it is more like an individually-autonomous community of people who "happen" to refer to their art with similar terminology, but undoubtedly more than one element seems to suggest the former option.

Within the vast context of popular music, it is particularly the electronic area to present the most interesting theoretical challenges. I shall therefore deal mostly with that, not forgetting, however, to refer to popular music as a whole, especially from a historical point of view.[18] The goal, evidently, is to create a connection between this topic and the general discussion on authenticity we are conducting in this section of the book. Instances range from such bands as Oval, who issued whole recordings based on the errors of a malfunctioning CD player,[19] to artists like the already-mentioned Matthew Herbert, who went as far as to write a "manifesto of mistakes" in music. Even when we listen to musicians not explicitly devoted to error aesthetics, we clearly understand that we are living a period of strong redefinition of the aesthetic phenomenon. Surely post-modernism is not the only period of such tension (I mean, *definitely* not. Marcel Duchamp, for one, would have something to say about that), but it is quite definitely an active and lively one.

Before any further reflection, it is maybe useful to provide some terminological clarifications. The English language (as many others) has several terms that are semantically similar to "error": "mistake", "fault", "failure", "blunder" "stumble", "fallacy", and others. Each of them carries precious analytical nuances, which—in many cases—are necessary for the purpose of the present essay.

The main problem, I feel, is to establish some limits, i.e. to circumscribe a given semantic area, wherein a discussion on error aesthetics is reasonably possible. Definition, history and theoretical analysis of the issue depend on this area. For instance, and I shall soon focus on this, a brief history of the musical error very much depends on this semantic differentiation.

Thus, let us try to define the word "error" with the help of a good dictionary. I took the liberty of making a short selection, among the very many definitions available, so that I can now transcribe seven—so to say—*autonomous* meanings:

1. An act, assertion, or belief that unintentionally deviates from what is correct, right, or true.
2. The condition of having incorrect or false knowledge.
3. The act or an instance of deviating from an accepted code of behavior.

---

[18]For Italian-speaking readers, I have dealt with this topic also in Martinelli (2010c).

[19]"[Oval's] music has been at the forefront of experimental electronica since the early 90s. [...] The group became known as the pioneers of the so-called 'glitch' movement. Since the group disdained conventional electronic instruments, their early albums were composed through the creative abuse of compact disc players. Deliberately mutilated discs were used to create a remarkable range of fragmentary sounds, which would then be painstakingly processed, compiled and assembled into musical pieces with a strange beauty of their own" (Inglis 2002). A remarkable example of this technique is Oval's album *Systemisch* (1997).

4. A wandering or deviation from the right course or standard; irregularity; mistake; inaccuracy; something made wrong or left wrong; as, an error in writing or in printing; a clerical error.
5. A departing or deviation from the truth; falsity; false notion; wrong opinion; mistake; misapprehension.
6. A moral offense; violation of duty; a sin or transgression; iniquity; fault.
7. A discrepancy between a computed, observed, or measured value or condition and the true, specified, or theoretically correct value or condition.

Apart from specific elements, what emerges here is a pragmatic classification of the error. In other words, that includes at least the following types: (a) **normative** (i.e. dealing with the violation of more or less juridical rules, that are socially and institutionally accepted); (b) **ethical** (i.e. in opposition with common sense and a general perception of truth and exactness within a given context or event); (c) **technical** (i.e. referred to what is considered the correct functioning and use of a given instrument or device); and finally (d) **structural** (i.e. dealing with the violation of the conventionally correct spatial-temporal setting of a given context or event). As always in these cases, these four typologies may easily overlap with each other. A normative error may very possibly carry an ethical character; a structural error may imply technical and normative aspects, and so on.

Moreover, another poignant aspect is the dualism presence/absence of **intentionality**, as referred to who or what makes the error. Which leads us to the dangerous area of logic. If I make a mistake on purpose, as with using an intentionally imperfect Remington font, can we really talk about error? Or—more subtly—if I "create the conditions" for an agent (say, a computer) to generate random errors, where do I stand on the chance-intentionality axis? Is chance the unavoidable element in a process that eventually produces an error? These questions are unfortunately not banal, since the inclusion or exclusion of several musical examples in the concept of error aesthetics depends on the answer we give here. As a matter of fact, the definitions I have selected, in their variety, suggest that both conditions are acceptable. Indeed, if it is true that in definition number 1 we talk about acts that "unintentionally" deviate from correctness or truth (I will not use these two terms as synonyms, I am sure I do not have to explain why); it is also true that a definition like number 6, stresses concepts like "offense" or "transgression", i.e. dialectical and behavioral processes that we normally think of as intentional, totally or partially.

We shall deduce that, when defining error in relation to the sender of the message, we have to accept both intentionality and chance, although the latter shall still be considered as more relevant, since it is the main element of novelty and opposition with traditional compositional conventions. Thus, we have to accept matter-of-factly the existence of a certain ambiguity between the two levels. Indeed, an error can be casual and intentional at the same time, since it can be provoked intentionally, and—so to say—"left free to develop" independently of any control. I may *decide* to throw a paint bucket over a canvas, but then I am most probably not able to *predict* the number and the size of spots that will be produced on my paintings.[20] In this sense, I feel it necessary to introduce a notion parallel to that of intentionality, which I shall call "**control**". If intentionality applies to the process of production of the erroneous message, control definitely focuses on its development during the stage of emission. The fusion of the two elements is what really originates the message, as perceived by the receiver.

There are also other criteria worthy of consideration. Firstly, I shall refer to the spatial/temporal dimension wherein the error occurs. If we think of errors in the traditional sense, as in the already mentioned case of typos, we normally think of something that is surely unintentional, but also episodic. Something that is likely to happen only sporadically, if I pay enough attention, and that is, actually, likely to become rarer and rarer once happened the first time (once written "typods" rather than "typos", I will probably pay more attention when writing that very word the next time). This, however, transcends the aesthetic discourse, or at least does not tell us the whole story. If I take the decision to use the error in an aesthetic fashion, then it is possible, but not at all normative, to preserve its **extemporaneity**. However, the aesthetic value of my artwork might be based on the **systematicity** of the error, up to an attempt to provide it with absolute continuity (as, musically speaking, in the case of loops). Both are strategic and stylistic choices: I certainly might be interested in the extemporaneity of the error, and particularly in that peculiar relation established with the specific portion of space/time wherein it occurs (aesthetically speaking, I might find the error appealing at point X of my song, rather than at point Y). But at the same time, I might be interested in the intrinsic musical quality of the error (i.e. I might just like "the way it sounds"), and so one should not be surprised if I wanted to reproduce it at more points of the song, or even in a continuative way.

Finally, we shall not avoid discussing the very particular relation established between the **error** and the **non-error**, that is, between our (provoked or not) violation of common sense, and common sense itself, i.e. a given phenomenon "the way it should be in ordinary conditions, without the error". Almost all words, in this latter sentence, should be treated with velvet gloves, as we are talking about so-called common sense, so-called normality, and so-called violation. In other words, if it makes sense to speak of "errors", it is because we are used to considering "exactness", and often "rightness" and "truth", as a different set of contexts and events.

Thus, the question is: what kind of relation does exist between wrong and right? What is the position of the error within that particular entity that qualifies it as error? I guess "antagonism" would be the answer of most people: an error goes "against" a non-error, somehow rejecting and neglecting it. In most cases, an error "spoils" the correct formulation of a given text. I shall call such a typology **anti-semantic** error,

---

[20]For the sake of my argument, I shall of course ignore that a Pollock or a DeKooning would hardly consider "error" the action of throwing a paint bucket on a canvas. The idea is that, in ordinary, traditional contexts, this action is a normative and/or technical error.

meaning the very probable chance that the error in question will threaten or ruin the correct sense of a message.

Evidently, though, this cannot be the end of the story. An error may spoil the sense of a message, the way it was thought, but it will not necessarily be meaningless. If I thought the sentence “John is a handsome cook”, we shall probably deduce that John is an attractive person who works in a restaurant, or simply an attractive person who likes cooking. However, if I made a mistake in writing the sentence, and wrote “John is a handsome rook”, then suddenly John would become either a nice specimen of *Corvus frugilegus*, or an attractive cheater in card games. In such a case, I shall name the error **hetero-semantic**: it has provided *other* meanings.

Now: an error is evidently more productive than a non-error. If the exactness of a given context corresponds to a limited number of meanings (restricted in many ways, starting from social conventions), its possible erroneousness opens unexpected doors to new meanings. If John, instead of being a handsome cook, was a handsome Cook, several interpretive options would be suggested by that capital C. Certainly, in comparison to the original formulation (that we here assume as “correct”), it was a mistake (what I had thought of was “cook”, not “Cook”), but now we might think that John is quite a peculiar cook, or even “the” cook by definition, or—alternatively—a member of a family named Cook. In other words, the error has helped the creative process (by offering new options). I shall call this error **neo-semantic**: it has *added* new possibilities to the same sign.

There is more. If, instead of being a cook, John was a “coo k”, I still would make a mistake (one too many space bars), but I am quite confident that most readers will get the meaning of the sentence anyway. “Coo k” is an expression that does not exist in English, but we cannot really define it anti-semantic, for after all it has not compromised the meaning of the sentence, as, instead, a cluster like “ckojohk” would probably do (in other words: the reader becomes aware that a typing mistake was made, but still understands that *I must have meant* “cook”). Thus, “coo k” must simply be considered an **extra-semantic** error: it goes “outside” the meaning, but does not necessarily contradict it.

Other errors have “parallel” functions to the message, as far as they do not threaten or change the meaning in the strict sense. However—e.g., by affecting the channel—they present the sentence in a totally new perspective, stimulating several interpretive processes in the field of connotation. “John is a handsome cook”, as written with the “Old Remington” font I mentioned, has certainly a different effect than the Times New Roman-written identical sentence. The meaning is not altered, but, for instance, all the little spots and smudges were not expected, and certainly they are not ‘correct’. Such an error shall be called **para-semantic**: it has provided an additional para-text that might (in the Remington case) place the sentence in a certain space/time context. All of a sudden, we might picture this John guy sometime in the 20th century, in pre-computer days, and—still travelling with imagination—we may imagine that he is a character appearing in some novelist’s manuscript.

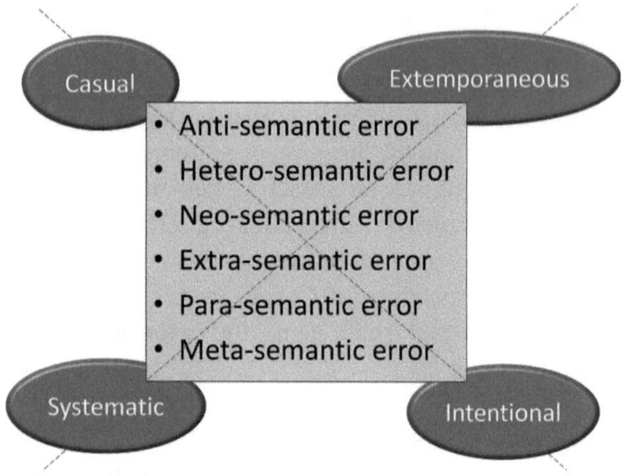

Fig. 3.4 Semiotic criteria and semantic relations for the notion of error

Finally, one cannot avoid facing a specifically semiotic dimension of the error, which—possibly more than any other—emphasizes the value of intentionality in many aesthetically-purposed errors. It is the case of the **meta-semantic** error, i.e. an error that actually interacts (by the usual means of transgression and deviation) with the text, in terms of complex and multiple articulations. If I wrote a sentence like *inverted commas John is a handsome cook inverted commas*, I would formally make another mistake, allegedly showing the incapacity to distinguish between lexical and punctuation signs. However, at this point, my aesthetic purposes are quite clear, since I could hardly pretend that such a mistake happened by chance. On the contrary, it appears that I am more concerned with the visual effect produced by the expression "inverted commas", than with the sign itself, in its ordinary function. It seems clear that *I did it on purpose*. I wanted to play with rules, through a little stylistic provocation.

No need to say, an error may easily contain more than one of the above-listed characteristics: in fact, that happens more frequently than such cases where just one single function is displayed. I hope that the specifically-musical examples I will later provide will make the theoretical picture clearer (Fig. 3.4).

3.2.3.1 A Brief History of Errors in Music

As the introduction of concepts like hetero- and para-semanticity, or dimensions like intentionality and control, seem to prove, my methodological preferences tend towards a rather 'wide' conception of error. This does not mean that the object of my analysis will be totally heterogeneous and omni-comprehensive, but surely—in doing so—it will be easier for me to trace a small history of the phenomenon.

At the same time, I should not take each of my reflections to the letter, otherwise I could easily end up confusing the concept of error in music with the much more familiar one of musical avant-garde. For instance, the contribution provided by

Satie and Debussy towards an emancipation of non-western scales within the western compositional process, is certainly relevant in a general sense, but it does not really make sense to qualify it as part of the emancipation of the error in particular, unless we gave a metaphorical and too wide a meaning to the word "error", in such a way that it would not stand as an autonomous concept anymore.

There is no room for a systematic history of the musical error, thus I shall describe, in a vague chronological order, some of the happenings in musical history that I consider relevant, if not decisive, for the formation of a keen artistic attitude towards errors. It shall soon be clear that error aesthetics and electronic music have a special connection: it was thanks to the latter—not only, but mostly—that a specific aesthetic area for the error to become *a value* was created.

An important figure is certainly that of Charles Edward Ives. In 1888, only 14 years old, Ives wrote a song entitled "Majority", which probably contains the first documented use of clusters in the piano accompaniment. Apart from being a disharmonic action by principle (thus, anti-semantic towards the concept of musical harmony), clusters can be considered an error and at the same time intentional (in Ives' case, but not necessarily in the case of a kid playing with a toy-keyboard), and uncontrolled, since their 'development' (the way they sound, the keys touched, the dynamic level...) is unpredictable and unique (one can hardly play two perfectly identical clusters).

The adoption of clusters had a large following already at the beginning of 20th century. It was Bartok, in his "Allegro Barbaro" (1911), who repeatedly used piano clusters over the composition. But it was mostly Henry Cowell (1912), then a young composer from the University of California, who extended and improved the use and the potential of clusters in musical composition. Cowell would use a whole arm and strong punches to produce clusters, extending thus the provocation to the field of performance too (and anticipating the likes of Jerry Lee Lewis in Rock'n'Roll of some fifty years). He was also a forerunner in the direct use of the inner piano strings, plucked and struck in a totally unusual manner. Cowell became one of the leading figures in American avant-garde, and his book *New Musical Resources* (1930, first written in 1919) became a 'bible' of the movement.

Thirty-five years after "Majority", Ives became protagonist of another form of intentional error, this time in the field of intonation: his *Three Quartertone Piano Pieces* (1923–1924) were indeed performed with two pianos tuned a quarter tone apart, thus producing a subtle but very clear effect of harmonic alienation. Both "Majority" and the Piano Pieces were excellent examples[21] of little but significant challenges to the musical conventions of the time (and of nowadays, to some extent): they were interesting and early forms of the aesthetic emancipation of the error (both events, although intentional, were iconic of unintentional circumstances).

Together with Ives' Piano Pieces, we shall also place Ferruccio Busoni's reflections on Thaddeus Cahill's Telharmonium. In his famous manifesto *Entwurf*

[21]They were more than two, in fact: Ives also experimented with polyrhythms and polytonality.

einer neuen Ästhetik der Tonkunst ("Sketch of a New Aesthetics of Music", 1907), Busoni celebrated the telharmonium as capable of producing microtonal scales.

Although the use of dissonances as an aesthetic element only partially pertains to the issue of error aesthetics, it is worth making some remarks on it. It is for instance the case of Bartok's *Twenty Hungarian Folksongs* (1906), whose employment of percussion instruments is largely anticipatory of what Stravinsky did several years later. Also Schönberg's dodecaphonic scales (first experimented in 1907) constitute an issue that is just partly pertinent to our scope, yet they deserve to be mentioned in terms of procedure. The production of dodecaphonic series, indeed, was for the Austrian composer the result of a combinatory mathematical process, whose structure went beyond the use of musical keys. Not that this should imply a specific connection, but many musicians interested in error aesthetics work in a very similar way with the computer: given a few variables (the twelve notes in the case of dodecaphony), the computer is required to generate a series of combinations. Once more, thus, we are talking about provoked-but-uncontrolled errors.

Serial music, a refinement of the work by Schönberg and his pupils Berg and Webern, achieved a large following during the 1950s, thanks to Boulez, Babbitt and Stockhausen: no need to say, the key-approach of these composers was precisely the random element.

Another relevant step that establishes an important connection between error aesthetics and the better known noise aesthetics is the coming of Futurism in 1909. Although mostly a visual art-related movement, Futurism gave fundamental inputs to music as well. In 1911, Francesco Pratella published a manifesto for Futurist Music that was strongly supportive of both Schönberg's work and the idea of extending harmony to the area of microtonal sounds. In addition, Pratella showed much interest in a freer conception of musical tempo.

One of Pratella's concerts gave inspiration to the futurist painter, Luigi Russolo, who wrote his own manifesto, entitled *The Art of Noise* (1913). In it, we find even more progressive ideas than Pratella's, who—as a matter of fact—was still tied to the use of traditional instruments. In contrast, Russolo wanted music to be emancipated from them: he proposed the inclusion of several noises within the concept of music ("It is necessary to break this restricted circle of pure sounds and conquer the infinite variety of noise-sounds", or—even more significantly—"we take greater pleasure in ideally combining the noises of trams, explosions of motors, trains and shouting crowds than in listening again, for example, to the Heroic or the Pastoral Symphonies"). Enthused by his ideas, Russolo decided to abandon paintings to concentrate on music only. Helped by another painter, Ugo Piatti, Russolo started to build machines named noise-intoners ("Intonarumori"), whose scope was to produce noises of the likes of roars, whistles, explosions, whispers, and so forth. Russolo formed a whole orchestra of intonarumori, and gave a concert in London in June 1914. The audience and critics' reaction was harsh: a journalist ironically begged Russolo to quit his career, after having heard "the pathetic cries of 'no more' from all parts of the auditorium". As almost all pioneers, thus, Russolo too was not understood by his contemporaries. Some time later, his machines were asked for by some theatre companies, in order to produce sound effects.

Definitely, Varese and Cage were active in more receptive times, but the audiences were still unprepared for noises in music when Erik Satie, in 1917, included sounds like steamship whistles, sirens, motor engines and other within the music written for the ballet *Parade*. Famous is Satie's response to one of the many critical attacks: "Sir and dear Friend, you are not only an arse, but an arse without music".

The inclusion of noise in musical aesthetics became later on quite a systematic, controlled and intentional attitude (see the case of Pierre Schaeffer and his musique concrète), and it should therefore be clear that noise aesthetics remains a relative, not a twin, of error aesthetics.

John Cage, unmistakably, deserves plenty of attention. Unlike many of his contemporaries, Cage would not aim at the total control of the compositional process, but was rather deeply fascinated by indeterminate and uncontrolled combinations of sounds. In 1950, Cage used "I Ching" as (casual) criterion for organizing sounds, in such a way that he was totally excluded from taking creative decisions. The same happened with his tape compositions: "William Mix" was a random assembly of hundreds of natural and artificial sounds. Cage's influence on posterity was very strong, especially in terms of redefinition of the threshold between musical and non-musical. In his book *Silence*, Cage wrote:

> It [An experimental action] is simply one the outcome of which is not foreseen. Being unforeseen, this action is not concerned with its excuse. Like the land, like the air, it needs none. A performance of a composition which is indeterminate of its performance is necessarily unique. It cannot be repeated. When performed for a second time, the outcome is other than it was. Nothing therefore is accomplished by such a performance, since that performance cannot be grasped as an object in time. A recording of such a work has no more value than a postcard; it provides a knowledge of something that happened, whereas the action was a non-knowledge of something that had not yet happened. (Cage 1961: 39)

What is even more relevant to the purpose of my arguments, is that Cage, in an interview given to Thomas B. Holmes (author of the excellent *Electronic and experimental music*), declared that the common factor in his works based on chance operations is the absence of intention, whether such actions are determinate or not (Holmes 1985: 130). If we translate this statement with the terminology I am using in the present essay, we have another confirmation of the difference and complementarities of the concepts of intention and control.

Years after Cage's experiments, something interesting happened within popular music as well. Towards the end of 1964, the Beatles were recording one of their many singles, "I feel fine". The song owns a couple of interesting records. It was one of the very first music videos in the modern sense of the term, and it probably marked the introduction of errors as a compositional source in popular music. Paul McCartney recalls the event as follows:

> John [Lennon] had a semi-acoustic Gibson guitar. It had a pick-up on it so it could be amplified... We were just about to walk away to listen to a take when John leaned his guitar against the amp... and it went, 'Nnnnnnwahhhhh!' And we went, 'What's that? Voodoo!' 'No, it's feedback.' 'Wow, it's a great sound!' George Martin was there so we said, 'Can we have that on the record?' 'Well, I suppose we could, we could edit it on the front.' It was

a found object – an accident caused by leaning the guitar against the amp. (McCartney, in Miles 1997: 172)

McCartney is obviously talking of that distorted bit preceding the riff, at the beginning of the song, and it is significant (in fact, very appropriate) that he is comparing this "artistic gesture" to Duchamp's *objet trouvé*, thus as the use of an *object* (an acoustic phenomenon, in this case) that is not normally considered art.

Similarly, even though I have no idea where to chronologically locate the first examples, a tendency to value "distraction" and "approximation" as stylistic choices became quite common. Keeping up with The Beatles as an example, this choice had already appeared in the album *Revolver* (1966: the coughs and hums at the beginning of "Taxman"), it developed very much in *The* Beatles, aka *The White Album* (1968: almost all songs are good examples, but certainly unforgettable is Ringo Starr's cry "I've got blisters on my fingers" at the end of "Helter Skelter"), and became programmatic in *Let it be* (1970: the album was born with the intention of creating an unpolished "back-to-basics" sound). Noises, background comments, little technical mistakes, hectic/uncertain endings: all these procedures were consciously (in many cases) or unconsciously (in a few others) used as metaphors of the character, the humanity and the warmth of rock music (indeed: the authenticity). Nowadays, and very often, musicians declare that certain little errors were left in the final editing of the record, in order to provide it with a feeling of realness and truth.[22] To borrow the words of Matthew Herbert, one of the electronic musicians mostly devoted to error aesthetics, "I am really into the idea of accidents at the moment. They provide the humanity in this sterile world. To paraphrase Don Delillo in *Underworld*, crowds (in sport) wait for something to happen between 'magic and accident' and it's true of most creative things" (http://www.magicandaccident.com/accidental/index.htm?2).

If we take the Beatles' *White Album* again as an example, we can compile an extensive database of the (casual or not) use of chance and errors in the musical composition:

- Songs that one should normally fade out are left free to "degenerate" with all their imperfections (e.g. "The Continuing Story of Bungalow Bill", with the final applause, as if it was a live performance);
- Elements that should simply *support* the performance, and that should finally be excluded from the final mix, are, however, put in the spotlight (e.g. the metronome in "Blackbird");
- Old imperfections that technologies had fixed are recovered as a precise stylistic choice (e.g. the gramophone-effect in "Honey Pie").

Etcetera, without forgetting the massive use of background noises and voices, and of course the Stockhausen*esque* "Revolution #9".

Apart from the previously mentioned implications, such procedures have a direct impact on the spatial/temporal dimension of the song. For instance, they put into

[22]Which is a funny paradox after all: the truth achieved through the error.

question the "official" structure and length. When does "Helter Skelter" end? When it fades out the first, the second time, or when the poor Ringo complains about his over-exploited fingers? And what about "Bungalow Bill" itself? Does it start when we hear the refrain "Hey, Bungalow Bill, what did you kill?", or a little earlier, when we hear that totally-out-of-context Spanish guitar bit[23]? We could say that the error makes the space/time coordinates quite flexible.

But there is more. The error often openly confuses the boundary between one song and another: does "Bungalow Bill" end with the "Hey ulp!" exclamation, or is that the beginning of "While My Guitar Gently Weeps"? This goes up to the extreme that a passage, introduced with the intention of seeming accidental (and left where it is because "it sounds good"), may risk becoming a musical entity in its own right, although in most of the cases anonymous. If the sleeve-notes do not report any song between "Cry Baby Cry" and Revolution #9, then what is that cute little tune that goes like "Can you take me back where I came from?" between the two mentioned songs? What if the tune is played on the radio? Do Lennon and McCartney get any royalty[24]?

What seems certain is that these accidents become a consistent part of the musical and cultural identity of the song in question. I challenge any beatlemaniac singer to declare that s/he never declaimed "'I dig a pigmy', by Charles Hawtrey and the Deaf Aids. Phase one, in which Doris gets her oats" before singing "Two of us" for his/her friends.[25]

Another historical key-topic consists of that long red line connecting the so-called garage-rock of the 1960s with most of the punk from the 1970s, and that stylistic notion (genre?) known as lo-fi since the end of the 1980s. It was then, indeed, that the expression "low fidelity" became not only a way to describe the recording quality of a given song or album, but a style on its own. Obviously (luckily, in a way), not all rock recordings had the chance of being performed and recorded with technologically high quality equipment. In many cases (like, indeed, garage, punk, etc.) a non-optimal recording quality was nothing more than just the result of a limited amount of technical and financial resources. Eventually, and exactly starting from the late 1980s, the term started to be associated with a specific area of independent musicians that would record on four-track machines. The American underground movement (represented by the likes of R.E.M.), some post-punk British bands, and New-Zealand acts like The Clean made of this apparent poverty a true trademark, applied to both traditional rock songs and more experimental compositions.

Coherently with this principle, independent labels began to issue lo-fi material mostly on vinyl, in a period (the one between the 1980s and 1990s) when it was the

[23]Which, incidentally, was a mellotron's preset, and not a real Spanish guitar.

[24]The practice of including apparently accidental songs has become very fashionable nowadays with the so-called ghost-tracks, placed at the very end of a CD, after a few minutes of silence from the official last song.

[25]For those unaware of Beatle anecdotes, that nonsense sentence was pronounced by John Lennon, and appears on the "Let it be" album, right before the beginning of "Two of us".

common opinion that CDs had surpassed LPs in every single technical aspect. The average listener started to become aware of lo-fi, as stylistic model, around the early 1990s, thanks to the emergence of the grunge scene, the so-called post-rock, and highly successful albums, such as U2's *Achtung Baby*. The paradox—or maybe not?—was that lo-fi sounds started to be created with the support of very sophisticated (often expensive) studio devices.

3.2.3.2 Errors in Music and Their Role

What benefit does a musician receive from errors? The above-mentioned Matthew Herbert went as far as to publish, on his own web-pages, a personal contract for musical composition that includes a so-called "manifesto of mistakes". I suppose it is worthwhile reading the whole of it:

> PERSONAL CONTRACT FOR THE COMPOSITION OF MUSIC [INCORPORATING THE MANIFESTO OF MISTAKES]
>
> THIS IS A GUIDE FOR MY OWN WORK AND NOT INTENDED AS THE CORRECT OR ONLY WAY TO WRITE MUSIC EITHER FOR MYSELF OR OTHERS.
>
> 1. The use of sounds that exist already is not allowed. Subject to article 2. In particular:
>
> A. No drum machines.
> B. All keyboard sounds must be edited in some way: no factory presets or pre-programmed patches are allowed.
>
> 2. Only sounds that are generated at the start of the compositional process or taken from the artist's own previously unused archive are available for sampling.
> 3. The sampling of other people's music is strictly forbidden.
> 4. No replication of traditional acoustic instruments is allowed where the financial and physical possibility of using the real ones exists.
> 5. The inclusion, development, propagation, existence, replication, acknowledgement, rights, patterns and beauty of what are commonly known as accidents, is encouraged. Furthermore, they have equal rights within the composition as deliberate, conscious, or premeditated compositional actions or decisions.
> 6. The mixing desk is not to be reset before the start of a new track in order to apply a random eq and fx setting across the new sounds. Once the ordering and recording of the music has begun, the desk may be used as normal.
> 7. All fx settings must be edited: no factory preset or pre-programmed patches are allowed.
> 8. Samples themselves are not to be truncated from the rear. Revealing parts of the recording are invariably stored there.
> 9. A notation of sounds used to be taken and made public.
> 10. A list of technical equipment used to be made public.
> 11. Optional: Remixes should be completed using only the sounds provided by the original artist including any packaging the media was provided in.
>
> (http://www.matthewherbert.com/pccom.php)

Obviously, it is point 5 that attracts our attention the most. Point 6 is interesting as well, and—more generally—points 1 and 2, for they seem to encourage a creative tension towards "magic and accident" to the detriment of the "pre-cooked-ready-to-serve". Point 5 is paradigmatic in several respects: it informs us that (a) the error is highly respectable *compositional material*, as useful as a non-error; (b) error is a hypertext, i.e. a clickable area from where a series of creative "windows" can be opened; and finally, however banal it may seem, (c) error is beautiful, i.e. it has an intrinsic aesthetic value, that is very rich in terms of unpredictability and heterogeneity. Point 6 seems to reinforce the value of randomness, as applied to the mixing phase. A combination of effects and equalizations that worked for song X, does not have to be removed a priori when working on song Y: who knows?—it could work there as well.

Herbert's contract is explicitly a small handbook for personal use, however it is extremely interesting, for it represents the point of view of a composer on the compositional process. And, as a matter of fact, is yet another confirmation of Cage's intuitions.

This is, thus, the first important advantage provided by the error: it is an incredibly rich compositional source, both in itself, and as a basis for further development. Another illuminating statement, in this sense, comes from Sebastian Oschatz, member of the already-mentioned Oval, another band with a preferential connection with error aesthetics:

> With the early Oval works we used random processes like sampling prepared CDs. We mostly prepared them without even having heard the material in normal playback mode. It was exactly like John Cage said: We never would have been able to come up with these samples by any means of inspiration or composition. These samples were not imaginable beforehand. Later I used a similar strategy with the videos: Starting with some designed elements, I introduced random elements by deploying certain bugs in the graphic software of the computer. These bugs usually had some very aesthetic appeal. I was working at a research institute on a very expensive silicon graphics machine at the time. I allowed myself some time to play with these artefacts and found some local maximae of the resulting aesthetics (unfortunately these kind of bugs seem to be sorted out on today's machines). The time was limited, because the institute where I worked closed at midnight. I recorded all material on two evenings. After that phase I got back into design mode and assembled the fragments to the music. This was a classic design task in a video editing suite. But the intermediary step was very important. Randomness is a very valuable tool, when economic pressures come into play. (http://www.meso.net/oschatz/oval/deutsch.html)

In other words, Oschatz suggests that errors and chance go beyond our imagination, for they act within areas that our perceptive organization, as a matter of fact, *excludes*, in order to better cope with the complexity of reality.

There is an interesting notion, in biosemiotics, named "dualism of the codes". The basic assumption is that the interaction living beings-environment consists of two important stages, the perceptive-behavioral one (which is of analogical type) and the mnemonic-organizational one (which is in contrast digital). Banally enough: when we walk, we of course act analogically, i.e. in a continuous manner. However, when we classify the act of walking in our mind, we tend to segment it in discreet units (as it was composed of separate moments, like "backwards", "forward", "slow", "fast", and so forth). Similar forms of cognition occur roughly in every moment of our life, and apply to the most diverse objects, events, people, groups, etc. This is totally understandable, since perceiving analogical information and processing it still analogically, is an impossible task. Not even computers do it: their enormous potential to store information is due to repeated acts of digitalization (bits), which can be so many to actually "resemble" analogical processes, but they are never so (the example of the increasing perfection and quantity of pixels is paradigmatic: a high resolution image can by now look like a real one, but it is still composed of pixels, that is, digital units).

As humans, too, we need to somehow digitalize information, decompose it, divide it into stages, rather than coping with a continuous flux of events. And, of course, the fewer the units, the easier the information management. Let us take politics as an example. The range of people's political opinions is of course an analogical whole of elements: everyone has his/her own opinion, the same opinion has different nuances, and so forth. A first necessary digitalization (in order for a community to be fairly represented) is the division of opinions into political groups, currents and ideologies. Then, we have parties: Communist Party, Radical Party, Neo-Fascist Party, Liberal Party, etc. Then we get to political tendencies: extreme left, left, centre-left, centre, etc. Finally we get to the poles: left and right, with the possible inclusion of centre (not to mention those people who classify politicians in the one category of "thieves").

In this process of continuous refinement of reality by approximation, a (more or less wide) margin for the unexpected and accidental is created. Our perceptive process reduces the endless variables in the surrounding environment, so, somehow, it leaves them aside as "irrelevant" entities. When these entities occur, they are not recognized: they are, exactly, errors. It is easy, thus, to guess how high the creative potential of these entities is: everything is fresh, new and rich. If anything, one risks only getting lost, but after all the risk is lowered by avoiding control. The error may be provoked, but is not usually controlled. The composer lets the error develop autonomously, and observes/admires the whole process taking directions as s/he would have never been able to predict through a digitalized thought.

Error aesthetics is thus part of a larger musical process. Not only do rigorous compositional rules take place in this process, but the point is that rules—so useful on the one hand—risk trapping the musical process. Herbert's manifesto is, to some extent, the representation of a quiet revolution. The musician does not refuse rules (see the other points in the contract), but decides that their ethical and technical opposite is important too. Paradoxically, the error will lose this potential when it becomes itself a dogma, a codified convention. But that is yet to happen.

Oschatz describes this quiet revolution very well:

> The main problem, the main advantage with randomness is that it at first negates the existence of an underlying structure, which could potentially be understood and deployed for some purpose. [...] Learning from random processes is another interesting thing, as Markus Popp [leader of Oval] did in the later Oval releases, or as it can be seen in the "clicks and cuts" school on [Oval's label] Mille Plateaux. The sensibility of working with rhythms and digital sound glitches is educated by the exposure to random sounds and structures. (http://www.meso.net/oval/deutsch.html)

Besides these ideological aspects, the error displays further interesting compositional sources. To start with, its special connection with electronic computer music shall be underlined:

> The accidental nature of not only the impetus of creation but also of the process itself is conceptualized in its purest form in the sampler. Since sampling with a microphone records not only sounds but also time and space, (every sound except silence has a pitch, an amplitude, a timbre and a duration) accidents are built into every sound sampled. You cannot go back in time and perfectly recreate the sound itself nor your recording process. In this instance, therefore, as well as your intended sound, sampling captures your physical environment including other people present, all additional sounds beyond the intended that occur simultaneously whilst recording, your position, your entire world in fact at that time and within an estimated and nominal radius of between 0.0000001cm and 10,000m depending on the sensitivity of the microphone and the amplitude of the peripheral sounds. This sound is then ordered, shaped and arranged into a bigger whole made up of other sounds with their own history and can be played as a pitched note in exactly the same way as any other electronically produced sound. This is such a huge concept: the idea that sound exists only once and contains stories within stories to be then reorganized, redefined and placed outside of their context to create and consider a new one continues to fascinate me. It makes the use of the sampler as a tool of theft of other people's music rather less enticing. The logical consequence of this process then is sounds within music with inherent meaning. (http://www.magicandaccident.com/accidental/index.htm?2)

More generally, it is evident that a semantically-limitless category as that of errors requires a pragmatically-almost-limitless system as a computer. As a matter of fact, to establish a compositional interaction with errors is easier when one is supported by devices that can codify them to different extents.

There is more. I already mentioned the capacity of errors to affect the spatial/temporal dimension of music. Using The Beatles as an example, we already explored some aspects of this phenomenon, but there are others of equally interesting semantic value. I shall mention two significant cases: (1) songs like Buffalo Daughter's "28 nuts", whose long introduction (actual part of the song) consists of guitars and other instruments in the act of tuning; and (2) songs like "Error System", by David Cunningham, featured in the album *Grey Scale*, whose sleeve notes describe the very peculiar principle the song is based on:

> Error System: The players play a repeated phrase. As soon as one player makes a mistake, that mistake is made the basis of his repetition unless it is modified by a further mistake. Thus, each player proceeds at his own rate to change the sound in an uncontrollable manner. On no account should "mistakes" be made deliberately to introduce a change into a performance. In short…sustain your errors. The water piece and the guitar piece are analogous to this process. However the process is automatic here, an inherent quality to the machinery used. The numerical sequence used in Ecuador and Bolivia is: 112211112233321122223311123322333311222322 repeated three times. The last time, an extra 11 is tagged onto the end. Each number refers to a bar, repetitions of a specific number are uniform. Notated according to the Ecuador system, Venezuela is 11223344556677 etc. for each player except the musician supplying the constant pulse. (http://www.markallencam.com/fldiscgreyscale.html)

Finally, I shall add few notes on the previously-mentioned character of "humanity" that error supposedly decorates music with. Such a character is not solely provided by random errors (that producers do not exclude from the final mixing). Other—definitely intentional—techniques are deliberately adopted against conventions (error as "the act or instance of deviating from an accepted code of behavior") in order to stress this particular realistic quality of sound. To mention just a few examples, some works by David Pajo, one of the most important figures in American post-rock (member—sometimes contemporarily—of bands like Slint, Aerial M, and Tortoise) are particularly intriguing. A song like "Rachmaninoff", included in the excellent album *Aerial M—As performed by Aerial M*, features two rather interesting characteristics: the tuning of the electric guitar (the sole instrument featured in the song) is mediocre (and I mean mediocre, not bad: it sounds as if it was tuned with some approximation, without a tuner); while the sound of the instrument is recorded in such a way that those particular whistles heard when fingers move on the guitar strings are very much in evidence, rather than hidden. Somehow despite its title, "Rachmaninoff" has an intentional garage-feel, and uses technical and technological approximation as a stylistic strategy.

It is also very interesting the use some singers make of their microphone (especially singers with a certain emotionally-strong impact in their repertoire, like the prototypical case of Tori Amos). The traditional distance between mouth and microphone is here violated, and drastically shortened, so that collateral sounds produced when singing (e.g. breathing) are recorded. The consequence is a strong sense of intimacy that contributes to the above-mentioned emotional impact (Tori Amos sings almost as if she were telling us a secret, or as if she were finding it hard to confess details of her life that are particularly sad).

In the end, it appears that errors are not as wrong as they might seem.

3.2.4 Conclusions

I shall conclude these few reflections with some specific remarks on authenticity as an ethical/ideological category, by far the trickiest of the problems related to the issue. Even a trivialization of the leading questions presents difficulties: is authenticity for progressives or for conservatives? For globalization or anti-globalization movements? Does authenticity promise a revolution (epistemological or even political) or a reaction? The funny thing is that everyone seems to have good reason to be a supporter/promoter/preacher of authenticity. And, inevitably, when an issue (especially a cultural one) is so transversal, the risks for a demagogical use of it are very serious.

I pointed out that, in order to discuss authenticity, one has to create *a case*, that is a relation between Oa and Ob, according to which, on the basis of given values and perceptions, Ob reminds us of Oa. Thus, what kind of case is authenticity? What is Oa, exactly?

A sign, however we define it, is a phenomenon of organic nature, even when we intend to use it in abstract and philosophical terms. Signs, trivially, are born, live and die. To die does not mean to disappear: it rather means, in common sense, to turn into something else, similarly as being born does not mean to be created out of nothing. The "death" of a sign corresponds, quite simply, to the point where a given number of vital sources, essential for its existence, cease to be active, "expire". Sources are time, of course, but also space or feeding (I mean feeding as a metaphor of anything that motivates the existence of a sign, like the presence of interpreters, who designate a signification for it).

Now, as an organic entity, a sign is by fact a *system* of signs (a difference exists, if it does, only at purely theoretical level). But, most of all, as an organic entity, and despite the changing processes occurring during its life (ageing, for instance, as in the case of the label "outdated" that is often applied to certain works or performances), the sign is still recognizable by one or more interpreters as *that* sign. Maybe, a bit older, with a few overused semiotic functions, but still, undeniably, that sign. However, when the sign actually dies (of course, I apologize for insisting on such metaphors, but I will soon get to the point) we have a real, true transformation. It is not like the sign disappears, or ceases to be perceived and interpreted. On the contrary. But it is evident to nearly everybody that it has undergone a clear, deep mutation, particularly in the direction of an organic decomposition (dispersion, division, occasionally explosion).

Such a process constitutes the very point when the sign can (or should) no longer be considered in its integrity, which simply is by now lacking. The closest thing to an idea of integrity that is left at this point is a series of representations of that integrity (obviously, the more the interpreters of that sign, i.e., the greater its semiotic relevance, the more these representations): memory, documentation, the signs of that sign (performances, remakes, re-arrangements, etc.).

But what happens to the various pieces of that sign? Of course, and still conforming to Lavoisier's rule, each piece is recycled in, or contributes to, the construction of another sign system. The decomposition of a musical work clearly contributes to other works, in terms of single traits, compositional strategies, performance techniques, etc.

At the same time, though, another phenomenon occurs, of a strictly cultural type[26]: the cult of the dead. The phenomenon of decomposition is mentally and socially managed in diverse ways, according to personal, religious, ideological and other inclinations, and only a strictly scientific attitude (in one way or another) succeeds in coping with the event in a fatalistic way. For instance, a typical case is the spiritual/religious tendency to believe that after death at least part of the sign will be immortal, therefore it makes sense to keep it alive in one or more of the following ways:

1. By believing in the existence of one or more transcendental aspects of that sign (therefore creating an ethical category to protect), aspects that are not empirically explicable, and that bring the sign itself into a dimension that is different from the tangible one (the latter admittedly impossible to achieve). The sign is thus transformed into an ideal (of beauty, purity, or something else).

[26]Thus, we are still dealing with a natural phenomenon, although admittedly a particular form of it.

2. Trivially, by trying to (metaphorically) re-animate the sign, typically hoping that, as some point, things will return to that condition, when the sign was alive. It is the transition from conservation to sheer reaction or resistance.

Such attitudes, predictably, emerge also in the area of the above-mentioned sign's representations. Memory, meant as (often absolute) value, sometimes becomes accurate documentation and, sometimes, active nostalgia. Sometimes both. In the light of these reflections, my (working) conclusion about authenticity (which I mostly intend as a form of intellectual provocation), is that the discussion around this issue is by fact:

1. A discussion around the state of health of Oa: Alive? Dead? Transformed? Decomposed? Immortal?
2. The ethical, ideological and sometimes demagogical assignment of one or more values to Oa, this latter becoming in most cases an ideal of truth, beauty, purity, etc.;
3. The ethical, ideological and demagogical assignment of one or more values to Ob, both as object in itself (whose aesthetic autonomy is now denied, not always with just reasons), and as related (i.e., opposed) to Oa;
4. The creation of a transcendental discourse category, while aiming to exist as a dimension fails to define itself in spatial-temporal terms becoming a (rather dangerous) meta-dimension. What and where are "the good old days"? If a synth-guitar is less authentic than an electric guitar, which is less authentic than a classical guitar, which is less authentic than the lute, etc., at which exact point of human evolution (human? And why not going even further back in time?), shall we reasonably locate the famous tradition to preserve and re-install? Needless to say, the suspicion is that this spatio-temporal location is a different one for everybody, and yet of the same type: it is the location of "familiarity", a condition in which the subject is at his/her ease with any stage of a sign's life, with both innovation and preservation. That condition is crystallized in the subject's mind as "ideal", and its loss, due to the natural course of events, is perceived as a deprivation, a robbery.
5. The assignment (usually arbitrary, although often reasonable in aesthetic and cultural terms) of more or less precise stylistic codes (or at least prototypes) to art, in a historical, geographical, social and strictly aesthetic sense. Such assignment creates one or more models (various Oa's), on the basis of which a case for authenticity becomes possible;
6. Most of the time (always?), such cases call for attention only when one or more Ob's appear. That is to say, using the previous metaphors, that the memory of the "dead" is somehow disturbed. Ob, in practice, creates a situation of conceptual "noise", opposed to a straight (musical, in our case) path that, because straight, was taken for granted;
7. This means that a case for authenticity is possible only on the basis of a claimed case for inauthenticity, or, forcing the comparison, that we remember the old only when the new appears, we remember simplicity when complexity appears,

or even in the most reactionary cases that we miss regression only when progress appears.

Having said that, and finally in defense of the concept, one shall also establish some fair distinctions. First of all, a critical position towards Ob (as inauthentic) may also be the result of pondered reflections, not only of a certain cultural laziness (as I clearly implied in my previous remarks). The rejection of Ob, that is, is not only aprioristic because it requires more mental effort than good old Oa. On the contrary, such rejection may be motivated by, for instance, the recognition of cultural and/or aesthetic dangers of trivialization of the above-mentioned stylistic codes. Ob is here recognized as inauthentic most of all because it makes the class of Oa more banal and silly, and also because, by imitating Oa, it is revealing a clear poverty of ideas and innovation.

Moreover, similar reflections neatly distinguish between a "forgery"-type of Ob, and an Ob that, on the other hand, seems to have constructively approached the stylistic lesson of Oa, resulting in an aesthetically valuable, if not autonomous, proposal. One certainly cannot speak of Berio's quotation of Mahler's 2nd symphony Scherzo, as a forgery, although it clearly is an Ob.

The question remains open: my attempt was most of all in the direction of defining the concept of authenticity within a numanistic perspective, in order not to get rid of it as a value, but rather to emancipate it from an ideological sacredness, which—more often than not—hinders the humanities' natural inclination for progressive and innovative research.

3.3 Case Study 2 (A and B): From Anthropocentrism to Posthumanism?

In this particular section, I intend to discuss two case studies under a single thematic umbrella. They both belong to a field of inquiry to which I have devoted a consistent portion of my academic career: that is, the humanistic approach to animal studies. Referring to the research map I showed at the beginning of the book, it is safe to say that animal studies are one of the most relevant entries there, particularly the semiotic approach to them. A field like zoosemiotics, for instance, is intrinsically "numanistic", as it contains at the same time a strong interdisciplinary emphasis (in the word itself: zoo + semiotics, that is, a biological subject within a humanistic field); a wish to use humanities (semiotics, in this case) as an added value for natural sciences (ethology and related fields, here); an attention to issues of current concern (the evolution of animal studies focusing on cognition have been rapidly developing in the last few decades); and—at least officially—an intention to base theoretical reflections empirically. Not by chance, I have had the opportunity to discuss the topics of this chapter on several occasions, most notably Martinelli (2010), so—just like the previous case study—I present here an upgrade and an update of previous research.

The relation between humanities and animal studies is becoming more and more relevant in modern research. Not too long ago, the New York Times commented on this increasing interest in a very insightful article (Gorman 2012). And, well, if a mass-medium feels it necessary to comment on an "academic trend", it means that the latter is all over the place and can no longer be ignored. Provided that the whole article is worth reading (also because it collects opinions from important representatives of animal studies such as Marc Bekoff, Dale Jamieson and Kari Weil), an insight to the incipit of the text may give an idea of the quasi-epic tones of the announcement:

> Once, animals at the university were the province of science. Rats ran through mazes in the psychology lab, cows mooed in the veterinary barns, the monkeys of neuroscience chattered in their cages. And on the dissecting tables of undergraduates, preserved frogs kept a deathly silence. On the other side of campus, in the seminar rooms and lecture halls of the liberal arts and social sciences, where monkey chow is never served and all the mazes are made of words, the attention of scholars was firmly fixed on humans. No longer. (Gorman 2012)

In the recent past, I have been member of the Academic Advisory Board of *Minding Animals*, an organization which is exactly part of this trend, and I was both proud and intimidated to share the board membership with some of my personal heroes in various fields: Peter Singer, Marc Bekoff himself, John Maxwell Coetzee and Jane Goodall. That is, a philosopher, an ethologist, a Nobel-prize writer, and a primatologist. This means that both scientists and humanists have welcomed this trend with sympathy, and authentic celebrities in these fields are ready to put their face to support it.

Yet, and arguably, there is no better example to illustrate the humanities' "allergy" to empirical evidence than seeing them in action when they discuss non-human animals, and particularly when they compare them with human beings. In this chapter, I intend to illustrate two case studies I have extensively worked with in the last ten years or so.

Also, within the economy of those new fields in humanities which—I feel—occupy a relevant place in the Numanistic forum, this chapter is meant to reflect the clear and strong connection between Numanities and post-humanism. The overcoming of anthropocentrism in the humanities is indeed one of the main entries in the posthumanist agenda:

> By the end of the 1990s (critical and cultural) posthumanism developed into a more philosophically focused inquiry (now referred to as philosophical posthumanism), in a comprehensive attempt to re-access each field of philosophical investigation through a newly gained awareness of the limits of previous anthropocentric and humanistic assumptions. Posthumanism is often defined as a posthumanism and a post-anthropocentrism: it is "post" to the concept of the human and to the historical occurrence of humanism, both based, as we have previously seen, on hierarchical social constructs and humancentric assumptions. Speciesism has turned into an integral aspect of the posthuman critical approach. The posthuman overcoming of human primacy, though, is not to be replaced with other types of primacies (such as the one of the machines). Posthumanism can be seen as a post-exclusivism: an empirical philosophy of mediation which offers a reconciliation of existence in its broadest significations. Posthumanism does

not employ any frontal dualism or antithesis, demystifying any ontological polarization through the postmodern practice of deconstruction. (Ferrando 2013: 29)

Both the premises and the conclusions of this chapter are perfectly aligned with this program (for more on the posthumanist agenda, see at least Halberstam and Livingston 1995; Badmington 2000; Miah 2008). Incidentally, it is my hope as editor in chief of this book series, to have the chance to host more than one reflection on these topics in our publications.

3.3.1 Prologue

Back in 2012, when my main academic affiliation was in Helsinki, I was invited to give one of those public lectures intended for a non-academic audience. For years, at Helsinki university, there has been this very interesting lecture series called *Studia Semiotica Generalia*, organized (as are the majority of semiotic events in Finland) by Eero Tarasti, the most prominent semiotician in that country. The lecture series, that year, had "humanism" as the main theme, and I was invited to give a speech on "animals and humanism". Now, I had two excellent reasons to decline that invitation, and yet I accepted. The first reason was that, at that time, I was having increasing difficulties in identifying myself with the contemporary semiotic scene in Helsinki, which in those years was exactly taking the very metaphysical drift that led to the current agony of semiotics in Finland: institutions became increasingly indifferent, when not irritated, by it, and it did not help that in 2010 a professorship and a master's program in semiotics were finally established in Helsinki after decades of efforts on Tarasti's part, but the competition saw a Peircean philosopher—instead of a true semiotician—as a winner. At that point, I believe, the capital sentence was written and signed: the master's program became soon an insipid variant of an ordinary program of philosophy, and at the same time Tarasti approached retirement, creating a perfect combination of excuses for Helsinki University to close that program even before the completion of its first 5-year cycle.

The second good reason to decline the invitation was the way I was asked to give this lecture. It was this very vague offer: "we are doing this lecture series on humanism… maybe you would like to say something about animals and humanism". Now, readers might not find this particularly funny, but the thing is, during my Helsinki years, I had developed this reputation of "Dario of the animals", and whenever any academic or semi-academic event about any topic was organized, I would get invited to say something about *animals and* that thing. Kind of like the old Tarzan novels: Tarzan and the Lost Empire, Tarzan and the Forbidden City, Tarzan and the Foreign Legion… so it was with me: animals and humanism, animals and Finnish national identity, animals and the seven dwarfs… you name it.

On the one hand, it was flattering, because it meant that people trusted my ability to be "pertinent" in nearly every context. On the other hand, I got this terrible suspicion that my colleagues were actually betting money on the possibility that I could actually come up with a suitable animal topic in whatever situation. I started having nightmares of this betting agency making odds: “Dario lecturing on Animals and Ice Hockey is 1 to 25”; “animals and the impact of boy-bands on popular culture is 1 to 30”… “Animals and humanism” was probably only 1 to 5, but still I might have made somebody a little richer on that occasion. Jokes aside, of course, the problem was the same as Sean Connery with James Bond: being cast in a role that prevents others to see you in your full potential as a researcher, thus becoming a *cliché* of yourself.

Why did I accept the invitation, then? I accepted because the two very reasons to refuse were also excellent reasons to accept. First, there and then, I thought it was important to convey the message that semiotics can still say something useful to interpret today’s society. Second, the fact was and is that, yes, animals *are* relevant in nearly any field of knowledge, because they are part of human life in many different forms. They are “animals”, to begin with, so they are living beings of the same kingdom we also belong to; but they are also philosophical concepts, anthropological archetypes, mythical projections, forms of discourse, and many other things. To accept the challenge of talking about animals and humanism meant, in fact, not to take any challenge at all, but simply to investigate a relation that had always existed.

And of course, on that very occasion, being that a *Studia Semiotica Generalia* lecture, I took semiotics as a case study. How does semiotics, as a “human” science, discuss the subject of animals, and is that a “humanist” discussion? I started by reminding the audience that humanism is a philosophical practice that espouses reason, ethics and justice as its main values. I asked them to store these three key-words (reason, ethics and justice) somewhere in their mind, and keep them as leit-motif of whatever would have followed in my lecture. I ask you now to do the same.

Have we noticed how many non-fiction books, on all possible topics except animals, actually start their arguments (especially in the introductions or in the prefaces), by writing that that very topic is something that “distinguishes us from animals”? One opens, say, an essay about politics, and there it goes “Human beings are political animals…”, or “unlike animals, human beings are political…”.

It is not whether these statements are right or wrong that is puzzling (although in most cases they are wrong). It is actually the *need* of saying it that is the peculiar part: the need of using a rhetoric opening of that kind, in a context where that comparison will not be mentioned again, simply because the book is not about animals, or about the human-animal relationship.

A little exercise that can be made is a basic survey on Google. Let us type something like “unlike animals, man” and see what we get (it is important to write “man” and not the politically correct “human beings”, because the latter is unfortunately still far from being a common formulation). So, we get two types of very interesting results:

1. As of late 2015 (while I am writing these lines), the entries on my computer are 2750, which is a lot, for a whole expression inserted with inverted commas (therefore excluding cases where not all the words are employed, or not in the same order).
2. Second, more importantly, one should actually check the contents of those entries. In an ideal world, an expression like this should be only used in natural sciences, when emphasizing (empirically proven) species-specific features of the *Homo sapiens* (as in "unlike other animals, man has such and such bodily function, skull size…"). On the contrary, what we get on Google is sentences like these (I shall copy and paste the first ten):

- Unlike animals, man is given a spirit in the likeness of God
- Unlike animals, man can deviate from his course of thinking and living however he sees fit
- Unlike animals, man does not merely exist within the narrow parameters of instinct
- Unlike animals, man was created in the image of God, that is, in the likeness of the Creator Himself
- Unlike animals, man has the "capacity of progress"
- Unlike animals, man is endowed with a greater number of instincts
- Unlike animals, man had a conscience and a religious instinct
- Unlike animals, man knows that he is different from other species
- Unlike animals, man fights his own species for other than directly sexual aims
- Unlike animals, man knows how to create objects and to appreciate them "in accordance with the laws of beauty"

And so forth. Basically, all sorts of things where an anthrozoological comparative analysis is either *not* called for or terribly outdated.

My last finding on the matter is a book called, of all topics, "The fundamentals of Interior Design". Interior design, ladies and gentlemen. Let us think about it for a second: we have to write a book on interior design, we are going to help people who want to furnish their own house, we are going to show a lot of nice pictures of houses with magnificent and expensive furniture… How do we want to start a book like this? Are we going to mention the history of interior design? Are we going to mention some famous brands or shops? Are we going to say that home is a sacred place, and therefore it has to feel cozy and beautiful?

None of this. We are going to write: "To be creative is part of the human psyche. It is one of those traits that set us apart from other animal species" (Dodsworth 2009).

Now, never mind that this statement is scientifically false: not only creativity in general, not only art, not only architecture, but even interior design specifically is something that other animals practice, and the case of the Satin bowerbird *Ptilonorhynchus violaceus* it is possibly the most outstanding. What is even more remarkable is why this gentleman felt the need to start his book like this. Did he have a purpose? Was he going to mention human uniqueness again, later in the book? No. He just needed a rhetorical way to kick off, and that sounded nice.

Anyway, we shall come back to this.[27] Now, a little step back is needed.

3.3.1.1 When Semiotics Deals with Other Animals

The main—and chronologically the first—semiotic field that deals with non-human animals is zoosemiotics. Zoosemiotics was introduced in 1963 by Thomas Albert Sebeok. That is the year when the term and a first definition make their first appearance, initially as a compromise between ethological and semiotic research (in the beginning, Sebeok was convinced that "zoosemiotics" had to be meant mostly as an umbrella term, gathering different scholarly approaches to animal communication—so, in a sense, it started with similar intentions as Numanities). A synthetic definition of zoosemiotics, in the light of its most recent developments, can be today that of *the study of semiosis within and across animal species. Semiosis*, following Charles Morris (e.g., 1946), is the process in which something is a sign to some organism. By consequence, zoosemiotics is interested in at least three important semiosic phenomena:

(a) The way animals make sense out of each other, or out of their environment: here, the emphasis of the process is on the receiver, who is the only subject surely taking part in the semiosis, and a true sender may be missing. I call this process *Signification* (not necessarily in agreement with the general trend of semiotics, which assigns it to "signification" a wider meaning—but this is of no concern here);
(b) On the opposite edge there is *Representation*, occurring when the sender is the only semiosic subject. In this case, zoosemiotics studies the way animals construct sense and, often but not always, *offer* it to somebody else; and
(c) *Communication*, occurring when sender and receiver both take part in the semiosic phenomenon, and therefore the above-mentioned "sense" (or text) is exchanged, understood or misunderstood.

This specification, besides its paradigmatic necessity, will also be needed later, because, to make my case, I have to distinctively refer to these three processes, and not just to communication (which is probably the only instance of semiosis that

---

[27]Not that posthumanism has improved so much this condition, by the way. The paradigmatic emphasis on the non-human as (finally!) an important cultural/scientific agent for humanities, is still counterbalanced by statements like the following (appearing in a book entitled *Inhuman Thoughts: Philosophical Explorations of Posthumanity*): "One ready response to this question is that humans do not engage in social and sexual behavior solely because of an evolutionary instinct for survival of the species. They also typically enjoy being social and being sexual. In the form of love of one's mate and of one's children, humans have combined the social and the sexual in a manner largely unduplicated among the other species. Unlike other animals, humans have historically refined these behaviours in numerous ways, influenced by circumstances and their desire for enjoyment" (Seidel 2008: 93). Oh, come on!

comes to mind, and was indeed—as we have seen—the primary area of inquiry identified by Sebeok). No: the research target of zoosemiotics, and in fact of the whole semiotics, is semiosis, the "action of signs".

This also means that another concept that is strictly related to the notion of communication, i.e., *intentionality*, is not necessarily part of a semiosic phenomenon. This also needs to be pointed out, in order to develop our discussion. A common tendency, which comes straight from the linguistic/semiological tradition, is that of considering only intentional messages as worthwhile of semiotic analysis. For semiotics to be *interested* in a given message, that message must be somehow *meant*. And this is why zoosemiotics is occasionally seen as a "threat" for (what it is considered to be) the real identity of semiotics. Such an identity—it is maintained—remains that of text analysis: it is acceptable that a painting, a building, a sonata are considered 'texts', along with literary works, but the alarm signals of vervet monkeys, the singing of blackbirds, or the dance of the honeybees cannot.

This stand, however, presents two main difficulties. On the one hand, to limit semiotics to the sole communicative/intentional dimension corresponds to label not just non-human semiosis as semiotically-uninteresting, but also most of the human semiosis as well: body-language, proxemics, interpersonal distance, plus several types of anthropological and social interactions. And this might be rather peculiar if one thinks that such a position is held by those semioticians who have a strong linguistic background, that is, they relate to a field that has a remarkable interest in non-verbal human semiosis.

On the other hand, and more importantly, a concept like 'intentionality' is not clearly-defined at all. What does it mean to *have an intention*? Is it the same as *to want* something, or is it something else? And how does one detect the existence of an intention in our thoughts/actions? In a famous article on primate deception, Byrne and Whiten (1991), traced a 0 to 3 scale of intentional behavior. The 0 level (observed when the animal reaches a goal in a completely random manner) is the only one that cannot be considered intentional. Level 1 (the simple aiming at a goal) is already an *intention.*

What seems to create the misunderstanding, in a commonsensical definition of "intention", is that, in human interaction, we witness the *awareness* and then the *verbalization* of the intention: we are aware that we want something and we can say it. This is a process that is often confused with intentionality itself, which—in turn—could be nothing else (or nothing more) than the main characteristic of the sign: referring to something else than themselves (Dennett 1996: 48). Whether this (reductive? Reductionist? We shall see that the phobia of "reductionism" is one of the main problems, here) definition is acceptable or not, it is clear that most of the time, when talking about intentionality, we tend to refer to the wrong phenomenon, i.e., the awareness of the intentionality, which is altogether—and evidently—a different matter. Intentionality itself seems to be a much more obscure concept. Peirce himself was rather generous in using the term semiosis in his writings, but at the same time wary of employing terms like communication and intentionality.

Switching the argument to a less serious mode, one could say that the object of semiotics is not only a sentence like "*I* am telling *you* to go from A to B, and *I mean it*", but also the more straightforward "Go from A to B".

Let us go back to the initial definition of zoosemiotics, now. What we suggested is that this discipline studies semiosis *within* and *across* animal species. This means that there is a range of semiosic phenomena that may be called "intraspecific", and another category that should be called "interspecific". By intraspecific, we shall mean the kind of semiosis occurring within one single animal species (or community, if we acknowledge the problematicity of the concept of species, to a certain extent—although we shall leave this issue for another discussion), i.e., within a group of animals that supposedly share a fairly similar perception of the world and similar ways to codify it. By interspecific, on the other hand, we shall mean the kind of semiosis occurring between different species (or communities), i.e., between groups that do not share the above-mentioned perception and codification of the world, if not to a very basic extent (this latter normally being the very ground for establishing a—temporary or not—common code). This, too, is a rather important distinction, because it implies a (sometimes radical) change of methodologies, and an address to a (sometimes radically) different order of problems.

Thirdly, the employment of the term "animal species" in the definition here provided is intended to cover the entire Animal Kingdom, i.e., the human species as well. This means not only that a part of human semiosic behavior (more or less, what transcends the linguistic domain, although the notion of language, too, is problematic, and that will be another important part of the problem) easily falls under the zoosemiotic domain, as ethology has already shown, but also that zoosemiotics investigates a field of knowledge that include both natural and cultural elements, and that—ultimately—the critical notion of "culture" is to be considered a part of the critical notion of "nature".

All considered, this has to be one of the most important, and brave, statements of zoosemiotics, for—among other things—it represents an attempt (which by now can be deemed fully successful) to extend the attention of semiotic research to the realm of the non-human, starting exactly from the assumption that a great number of characteristics that *we thought* were typical of human semiosis, are in fact to be widely reconsidered. Please, dear reader, keep this in mind, because it is a crucial part of the story.

Throughout almost its entire history, indeed, semiotics has always been an anthropocentric and logocentric discipline, with an exclusive emphasis on human- and language-related issues (this despite the fact that the earliest conscious examples of semiotics consisted in the medical observation of the body—symptomatology, diagnostics, etc.—carried out by the likes of Hippocrates or Galen of Pergamon). John Locke, in the 17th century, used the word "semiotics" to describe the "doctrine of signs". It was once again a human-centered enterprise. Yet,

> While his prime concern was with those signs of our ideas "which men have found most convenient, and therefore generally make use of," that is, "articulate sounds" or verbal signs, Locke was fully aware that other creatures, such as birds, also have perception,

> "retain ideas in their memories, and use them for patterns", in brief, that they are comparably served by signs. (Sebeok 1990: 37)

The real turning point, in the economy of what will later become zoosemiotics, occurred in the 19th century, as Charles S. Peirce acknowledged the semiotic nature of the non-human world (to him, the sign was a connective element not only in all experience and thought, but in the whole universe), but it was not until the biologist Jakob von Uexküll that the first, important, specific argumentation in support of another field appears: that field is nowadays known as biosemiotics: the study of semiosis in all living forms, therefore a broader area of investigation, which *contains* zoosemiosis. Uexküll's *Institut für Umweltforschung*, founded in 1926 at Hamburg University, investigated the perceptive environment of animals (i.e., their Umwelt[28]). Though not a semiotician, and probably never intending to be one, Uexküll brought to attention a number of topics of fundamental (bio)semiotic interest, and later his son Thure, and Thomas A. Sebeok, introduced his work to the semiotic community, labeling the German biologist with the infamous term "cryptosemiotician" (a fate shared with nearly all great contributors to human knowledge, at one moment or another of the ever-in-progress construction of semiotic history).

All this was going on while the then-dominant school of semiotics—or "semiology", as it was called back then[29]—of the Saussurean tradition, made it very clear that the discipline was a natural continuation of Linguistics, or even—as Barthes had put it—just a part of it. In Sebeok's account:

> In an independent but parallel tradition, amplified by F. de Saussure's heritage, semiotics, alias semiologie, has remained steadfastly anthropocentric, intertwined with language, le patron general of Saussure's programmatic science. Many linguists later tended to more or

[28]"Umwelt" means in German "Environment", but in Jakob von Uexküll's formulation came to designate a much more complex notion, usually referred to as "subjective universe" (Sharov 2001: 211), or also "semiotic world of organisms" (Kull 1998: 304). The starting assumption is that the environment inhabited by an organism is not merely the actual environmental niche, but is a larger not purely physical 'environment', of which the niche is just a part, that is perceivable and meaningful in its entirety only from the perspective of that particular organism. Uexküll (1982: 29–30) exemplifies this concept by describing the completely different meanings that a flower may have to a young girl (an element of decoration), an ant (a path for reaching food), a cicada-larva (construction material), and a cow (fodder).

More specifically, an Umwelt is the result of a *Merkwelt*, i.e., the specific perceptive field of a given organism, and a *Wirkwelt*, i.e., the field of actual interaction between the organism and the environment (the latter being also the organism itself, as the example of proprioception paradigmatically explains). Perceptual and operational elements come to establish the specific Umwelt of the given organism, which is exclusive for each species, in particular, but also—broadening the concept—for each community, individual, class, family and so forth. The Merkwelt and the Wirkwelt are constantly in action, as the organism (also called a "structure", or a "receiver of meaning") affects and is affected by the environment (the "counter-structure", or "carrier of meaning"). In that sense, the process is described as an "Umwelt circle".

[29]Sometimes the word "semiology", as opposed to "semiotics" and not as its synonym, is employed to mark the difference between the continental European school (semiology) and the American one (semiotics).

> less agree: thus L. Bloomfield asserted that "Linguistics is the chief contributor to semiotic," and U. Weinreich called natural languages "the semiotic phenomenon par excellence." But it was the prominent French critic Roland Barthes who—like W. H. Auden's "linguist who is never at home in Nature's grammar"—carried this glottocentricity to its preposterous (but perhaps playfully conceived) conclusion by turning Saussure's formulation topsy-turvy with his declaration that "linguistics is not a part of the general science of signs, even a privileged part, it is semiology which is a part of linguistics…" The validity of this paradoxical inversion of the customary order of things can be contemplated only, if at all, at the price of throwing all of comparative semiotics overboard by dividing the animate world into two unequal classes—speechless vs. language-endowed—and then consigning the sign behavior of well over two million extant species of animals beyond the semiotic pale. (Sebeok 1990: 38)

Back on the 'opposite front', the Peircean one, other signals of an upcoming new field of inquiry came from Charles Morris, the truest follower of Peirce, and from the oncologist Giorgio Prodi, who termed the study of biological codes "Nature Semiotics", and from Rothschild (1962), who first actually used the term "biosemiotics" in a scientific context:

> This approach presupposes acceptance of our position that the history of subjectivity does not start with man, but that the human spirit was preceded by many preliminary stages in the evolution of animals. The symbol theory of psychophysical relation bridges the gulf between these disparate avenues of research and unites their methods under the name of **biosemiotic**. We speak of biophysics and biochemistry whenever methods used in the chemistry and physics of lifeless matter are applied to material structures and processes created by life. In analogy we use the term biosemiotic. It means a theory and its methods which follows the model of the semiotic of language. It investigates the communication processes of life that convey meaning in analogy to language. (Rothschild 1962: 777)

One year later, as already mentioned, Sebeok coined the term and developed the theoretical paradigm of a specific biosemiotic field named "zoosemiotics", somehow inaugurating a new phase for semiotic history, a phase in which non-human semiotics is no longer ignored or underrated. This event certainly represents *the* milestone in zoosemiotic history, and shall be deepened later on.

Sebeok mantained that

> The process of message exchanges, or semiosis, is an indispensable characteristic of all terrestrial life forms. It is this capacity for containing, replicating, and expressing messages, of extracting their signification, that, in fact, distinguishes them more from the nonliving—except for human agents, such as computers or robots, that can be programmed to simulate communication—than any other traits often cited. The study of the twin processes of communication and signification can be regarded as ultimately a branch of the life science, or as belonging in large part to nature, in some part to culture, which is, of course, also a part of nature. (Sebeok 1991: 22)

Later, he added that "the life science and the sign science thus mutually imply one another". (Sebeok 1994: 114).

These reflections introduce a number of important key-terms and concepts that should be considered the pillars of the bio- and zoosemiotic disciplines:

1. The concept of semiosis, i.e., the action of signs, is the real target of semiotics;
2. All life forms are semiosic. Thus, semiosis is primarily what distinguishes life from non-life;
3. Culture and Nature are not concepts in opposition, but in fact the former is part of the latter;
4. If life science intersects with sign science, then semiosphere and biosphere are probably synonyms.

If we now had a Numanities-sensitive device, we could switch on the "Metaphysical drift-detector" function and probably by now a yellow light would be blinking. Not yet a red one, but we are getting closer. This would be not because there is anything metaphysical yet, but because—by enlarging the field to such an extension that, for instance, semiosphere and biosphere are the same thing—we are setting the preconditions for more adventurous and anti-empirical speculations.

In addition to these reflections, Jesper Hoffmeyer pointed out the centrality of semiosis in biological studies. To Hoffmeyer, the biggest contribution that biosemiotics can make to the life sciences is the emancipation of sign and semiosis as the crucial elements in life: semiosis is the "most pronounced feature of organic evolution", and signs are the "basic units for studying life".

> The most pronounced feature of organic evolution is not the creation of a multiplicity of amazing morphological structures, but the general expansion of 'semiotic freedom', that is to say the increase in richness or 'depth' of meaning that can be communicated. (Hoffmeyer 1996: 61)

And: "The sign rather than the molecule is the basic unit for studying life" (Hoffmeyer 1995: 369).

This position is somehow antagonistic to that held by Marcello Barbieri, who is on the contrary a supporter of a code-based biosemiotics, an approach that focuses on three important aspects:

> One is the idea that the cell is a duality of genotype and phenotype, i.e., a biological computer made of genetic software and protein hardware. The crucial point is that a computer contains codes but is not a semiotic system because its codes come from a codemaker, which is outside the system.
>
> The second basic concept is the idea that all biological novelties are generated by natural selection, i.e., by an agent, which is outside the cell just as the human mind is outside the computer. But if the cell is a biological computer assembled by natural selection, it is perfectly legitimate to say that it is not a semiotic system, and this justifies Florkin's statement that there is no real meaning in it. Ultimately, that leads to the physicalist thesis that there is no real code either at the molecular level, and that molecular semiosis is merely an illusion. The computer model of the cell, in short, keeps semiosis out of the cell, and this is why the first true model of molecular semiosis was the idea that every cell is a trinity of genotype, phenotype, and ribotype, i.e., the idea that the cell contains an internal codemaker [...] This was complemented by the idea that coding is not reducible to copying, and, therefore, that natural selection (based on copying) and natural conventions (based on coding) are two distinct mechanisms of evolution [...]

> Another important contribution to code-based biosemiotics came from the discovery of an increasing number of organic codes. That development started with the unveiling of the sequence codes by Trifonov [...] and has grown slowly but steadily ever since [...]
>
> The 'code based' approach to biosemiotics, in short, is a road that started with the recognition of semiosis at the molecular level and worked its way up by extending the concepts of code and meaning to the higher levels of biological organization. At about the same time, however, there was also another road to biosemiotics that was being developed. A road that went exactly the other way round, i.e., that started at the higher levels and worked its way down towards the lower ones. (Barbieri 2008: 594)

This discussion is one of the hottest within biosemiotics, and—if the people involved do not mind me gossiping a bit—it has also led to personal confrontations, letters of resignation from given scientific boards of given associations and journals, "diasporas" of scholars from one community/project to another, and other similar events that proves the extreme problematicness of this discussion.

Wanting to generalize, the notion of sign-based biosemiotics is more sympathetic to an idea of the field as a humanistic-oriented one, while code-based biosemiotics heads more clearly in the direction of natural empirical sciences. Zoosemiotics, in turn, is certainly eager to find a compromise between humanities and natural sciences in a paradigm that hopefully takes the best from both, yet one perceives that this compromise, on the part of a discipline like, exactly, zoosemiotics (i.e., a branch of "semiotics"), requires more effort in the direction of biology, than the other way round. In that sense, there *should* be (emphasis on "should") a natural sympathy towards any contribution that will help in keeping the discipline within the necessary empirical premises, and will indeed prevent any risk of metaphysical drift (a sympathy that will be more explicit later in this chapter).

In any case, both Barbieri's and Hoffmeyer's schools make a strong case for the centrality of semiosis in biological processes, and for the intimately interdisciplinary nature of biosemiotic research. As Barbieri himself states:

> There have been historical disputes between the two versions but [...] they are not incompatible, and both share the idea that every living creature is a semiotic system, i.e., that semiosis (the production of signs) is fundamental to life. (Barbieri 2008: 577)

3.3.1.2 Thinking Too Big?

Additional signals of preoccupation, for our numanistic perspective, appear when we proceed into the systematics of biosemiotics. Life semiosis, in general, can take place either *within* a living being (endosemiosis) and/or *between* two or more of them (exosemiosis). Endosemiotics, thus (or—according to the specific cases—protosemiotics, microsemiotics, cytosemiotics, etc.) studies the message exchange among cellular organelles, cells, tissues, organs and organ systems. When Hippocrates, for instance, analyzed the symptoms of a given disease as signs of the disease itself, he was basically pioneering endosemiotics.

Endo—is a Greek preposition that stands for *Inside*: the internal semiosis that takes place in an organism's body is extremely complex and is probably the most intense semiosic activity one may find. A single human body consists of about 25 trillions cells, a number which—alone—is 2000 times more than the entire human population on this planet. Plus, all these cells have direct or indirect connections with each other through more than one modality. The number of messages exchanged inside one single living organism is unimaginably high:

> Millions of so-called receptors capable of recognising specific signal molecules in the cell environment are located in the membranes of each of our cells. These receptors function as communication channels through which our cells, tissues and organs are persistently communicating with each other all around the body. Especially interesting is the recent discovery that receptors on the surface of immune cells are capable of decoding the messages exchanged among nerve cells and vice versa. The psycho-somatic integration of the nervous system, the immune system and the endocrinological system in a healthy organism is the result of this gigantic semiotic interaction among many thousand billions of cells, each of which is capable of interpreting a limited range of molecular signs. Disease may be seen then as the result of erroneous communication among our body parts. We fall ill because our cells cannot quite succeed in uniting to create us. (Hoffmeyer, in Bouissac 1998: 84)

Probably, the earliest forms of semiosis in our bio/semiosphere are those of the *prokaryotes*, one-celled nucleus-less micro-organisms, better known as bacteria.

Exosemiotics, in turn, regards the entire spectrum of message exchange between two or more complex organisms. Fields of exosemiotics are mainly *phytosemiotics* (whose object is the semiosis among plants), *micosemiotics* (semiosis among fungi) and, finally, *zoosemiotics* (semiosis among animals). A relevant branch of zoosemiotics is obviously *anthroposemiotics*, i.e., the semiosis among the human animals. Some people still consider anthroposemiotics as a separate field, to be lined alongside the other three, claiming that the distinctive feature is the presence of culture in anthroposemiotics only, but the classification is scientifically incorrect (the human being *is* an animal, not a distinct entity).

So far, nothing wrong, but we can already see that what we are dealing with, here, is a globalistic conception of life as a whole, that is, the proposition that the nature of life is intimately. Such stand, mostly promoted by the American semiotic tradition (Peirce and Sebeok first of all), led a few biosemioticians to the slightly pretentious belief that the entire Universe can be analyzed and interpreted semiotically:

> In its most radical version biosemiotics sees itself as "general semiotics", while traditional semiotics studying human sign systems is seen as just a special part hereof, anthroposemiotics. This understanding may eventually be coupled to a cosmological vision of evolution as a general tendency of our universe to strengthen the autonomy of the semiotic sphere relative to the physical sphere on which it depends. In system Earth this might further be seen as a trend in organic evolution towards the formation of species with increasingly sophisticated umwelts, or in other words towards a general growth of semiotic freedom, a trend which has reached its temporarily richest expression in the art, religion and science of human cultures. (Hoffmeyer, in Bouissac 1998: 85)

This is when our metaphysical drift detector turns to red. Let us read another definition of biosemiotics, this time provided by Kalevi Kull:

> Biosemiotics can be defined as the science of signs in living systems. A principal and distinctive characteristic of semiotic biology lays in the understanding that in living, entities do not interact like mechanical bodies, but rather as messages, the pieces of text. This means that the whole determinism is of another type. [...] The phenomena of recognition, memory, categorization, mimicry, learning, communication are thus among those of interest for biosemiotic research, together with the analysis of the application of the tools and notions of semiotics (text, translation, interpretation, semiosis, types of sign, meaning) in the biological realm. (Kull 1999: 386)

Apart from those notions that have already emerged from the previous definitions, Kull is here suggesting that the introduction of semiotics in biological studies is not only worthwhile in the direction of a new theoretical paradigm (e.g., the centrality of semiosis rather than morphology), but also in strictly methodological terms: Nature, in other words, can be read as a text, it can be interpreted, it has a meaning, and so forth.

So, is biosemiotics, after all, about just "everything"?

When a perspective of this type comes to mind, it is difficult not to feel a bit dizzy, and that is why we need to point out the dangers of, so to say, indulging too much in a promised land of "answers to all questions". It is a risk that other scientific communities have already perceived, and this is perhaps why the reputation of semiotics is not so high within the academic world at the moment, after having enjoyed a golden age of respect and admiration back in the 1960s and 1970s (i.e., when semiotics did *not* claim to have such an omnicomprehensive range of interests). And it is difficult to argue against that, when biosemiotic essays are called things like *The Physics and Metaphysics of Biosemiotics*; or *Biosemiotics as a Mode of Thermodynamics in Second Person Description*; or *Beyond bioinformatics: can similarity be measured in the digital world?*

Especially the theoretical foundations of biosemiotics, more than those of zoosemiotics, have been exposed to significant criticism, from both semiotic insiders and outsiders (typically from other biological sciences). According to Hoffmeyer,

> 20th century life sciences have been characterised by two major trends. One trend is molecular and genetic reductionism. This trend is well known and need no further comment. Beginning as an undercurrent to this trend, however, another much less noticed but in the long run just as important trend has gradually been unfolding: The semiotisation of nature. (Hoffmeyer 1997: 355)

It can safely be said that the second trend mentioned by Hoffmeyer was the one that made biologists most uncomfortable. The history of modern biosemiotics has also been a history of contrasts and subtle fights between semioticians and scholars belonging to other fields. To be fair, it should be said that this was due both to a conservative attitude of the traditional biological sciences, typically hostile to theoretical/methodological changes in their field, and to a certain occasional arrogance from biosemioticians, firmly convinced of the undisputable superiority of semiotics over other disciplines. To mention one example, Konrad Lorenz was definitely inspired by the work of Uexküll, and certainly ethology can be credited with having an intimately semiotic nature (its main branches being animal communication and sociobiology). But pushing that link as far as to say, like Sebeok did, that ethology is 'hardly more than a special case of diachronic semiotics' (Sebeok 1976: 156) is a tad too reductive towards a field which not only may easily have a synchronic approach too, but—more importantly—focuses its investigation on topics that all in all are not of great semiotic interest at all.

Statements like this even managed to irritate insiders of the semiotic community. Kristian Bankov, for instance, became so annoyed by this unlimited extension of the biosemiotic field, to actually conceive, in 2005, a sort of Ockham's Razor (ironically baptized Bankov's Razor by Bankov himself) to revise, among other things, "(1) the philosophic grounds of the biosemiotic discourse" and "(2) the scientific output of biosemiotics" (Martinelli and Bankov 2008: 399):

> Concerning the first point, one must distinguish between several interrelated discursive orientations within the field of biosemiotics. 1) A strong ontological pretense in biosemiotics, a philosophically-grounded claim on how things hang together, what the real essence of nature and life is, etc. "Life is coextensive with semiosis" is characteristic of that kind of reflections. 2) A huge part in biosemiotics, dedicated to a kind of history of ideas, relevant to its predecessors or discovery of biosemiotic ideas in "unexpected" thinkers and scientists. That includes contributions of systematization of the contemporary advances in the field. 3) A descriptive, strictly scientific, orientation, concerning facts in the Universe and the biosphere: Big Bang, cells, molecules, plants, bacteria, animal behavior, etc. The above-mentioned theoretical mimicry can be found mostly in those contributions. 4) Last but not least, the contribution on current socio-political issues like ecological consciousness, "biomoral" and others, where biosemioticians are expert voices in the public space.
>
> Considering this variety of discursive orientations, it is difficult to establish any clear identity for biosemiotics. After hearing that biosemiotics is "the study of signs, of communication, and of information in living organisms" (as stated in the Oxford Dictionary of Biochemistry and Molecular Biology) and even more: "the scientific study of biosemiosis" [...], an unprejudiced observer would imagine people with an experimental agenda. Kalevi Kull claims that "Biosemiotics means biology" [...]. My expectations were similar, but when I started reading books and papers I found much more philosophy than laboratory/field work. Somehow, the identity of the discipline was to be established more as an interpretive than as a productive approach. And I think that more or less all scholars in biosemiotics would agree on the precept, present in any extended definition, that biosemiotics offers the existing biological theories a kind of generalizing view which they lack and they need. (Martinelli and Bankov 2008: 399–400)

The notion had the great merit of shaking the coconut tree of biosemiotics, forcing some insiders of the field to seriously reflect upon the scientific role of biosemiotics, its weaknesses and its strengths. There was quite a hype around the Razor in the 2–3 years following its enunciation, within the semiotic circle. In one the main points of his argument, Bankov notes that a certain "scientific humility" that he expects from biosemiotics is, "not present in the majority of biosemiotic papers I have read" (Martinelli and Bankov 2008: 401).

3.3.2 *Case Study 2A: (the Trouble with) Human Special Specificity*

If this identification of the concept of "life" with nearly everything brings biosemiotics on the (often trespassed) threshold of metaphysical drift, both biosemiotics and zoosemiotics display more subtle cases to which I would like to devote my attention from now on. To discuss metaphysical speculation as a case of humanistic research that forgets entirely empiricism is certainly important, but in a way all too easy. It allows us to emphasize a situation that could be emphasized anyway, without much effort (and certainly without the need to develop a brand new forum, with a new name and a book series attached to it). On the one hand, indeed, we have this all-inclusive mentality celebrating Peirce's famous statement "The universe is perfused with signs" in all possible forms; on the other hand, we notice an interesting form of *horror vacui* from the opposite side of the discussion: the human subject is carefully kept outside the discussion (meaning that so far no biosemiotic paper/article—not that I know of, at least—focuses on human-related topics). The obsessive recruitment of every single organism, atom, celestial body, cell in the biosemiotic (and partly in the zoosemiotic) program is counterbalanced by an equally meticulous exclusion of anything "human" from the discourse. Why so? Is *Homo sapiens* not a living form?

The thesis I intend to defend here is that the motivations behind this are more ideological than methodological, and I shall like to demonstrate this by analyzing the way (bio/zoo) semiotics has tackled the topic that sooner or later every human field of inquiry feels the need to tackle: a definition of "humanity" in qualitative (discontinuous) distinction from the rest of the animals. Such definitions have a long history that go back to the dawn of history, and always present a distinctive, altogether amusing, pattern in three parts: the same goal (founding human identity by establishing a radical difference with all the other animals), the same action (launching a manifesto-statement on the specific and special diversity of the human being within—and often without—the Animal Kingdom), the same methodology (trial and error). Only the aftermath present two variants: (1) as soon as the statement is dismissed by a scientific counterproof, another statement, concerning another supposedly species-specific human characteristic is issued; or, alternatively, (2) if the characteristic in question is perceived as possessing strong potentials of human exclusivity, there is the option of narrowing the definition of the characteristic itself, until, fatally, other animals are excluded. Of course, in the latter case, several human communities or subjects are excluded too, producing what in military jargon is called "friendly fire". In fact, warfare vocabulary is quite helpful here, as "struggle" is possibly the most appropriate word to describe this process: a struggle for affirming the "ingroup" identity, in a specific historic moment where the only morally-acceptable discrimination can be operated on the outgroup "animals".[30] We shall later deepen this particular aspect.

There is no pretension here to reconstruct a philological chronology of such statements, but at least the following should be mentioned. At one moment or another of history, the human being ("man") has been defined as:

- The rational animal
- The symbolic animal
- The playful animal
- The moral animal
- The laughing animal
- The spiritual animal
- The social animal
- The cultural animal
- The linguistic animal
- The abstract animal
- The signifying animal
- The political animal
- The tool-maker animal

Some of these definitions stem from authoritative thinkers, who, in a certain moment of philosophical history, reflected on the characteristics of human nature (the notion of rational animal comes from Aristotle, that of the symbolic animal is best known in Cassirer's formulation, and so forth). Some others derive from recent studies in sociology, psychology or else, and some others, finally, are of a commonsensical or superstitious/religious nature.

Now, as resulting from different investigations (mostly in natural sciences, but occasionally also in humanities, including semiotics[31]), all these supposedly distinctive features of humanity were challenged and generally invalidated exactly in their uniqueness. Any of these definitions would last a certain amount of time as the "ultimate sign of human distinction", and then the likes of Darwin, von Frisch, Griffin, Goodall, Sebeok and others would raise their hand and go like "well, actually, we found out that also animals can do this". Ever relentless, other scholars would then put their efforts into finding *another* sign of human distinction, eternally convinced of the importance of the enterprise. Or, as we said, if it is not "another sign", then it is the same sign, but heavily narrowed in its definition, with the purpose of creating such a small circumscription of the concept to finally keep other animals out. It happened with "culture", "language", "mind", "abstraction", "art", and other critical topics, whose definition and conceptualization are in constant progress, and therefore very liable to be re-discussed (i.e., in most cases, restricted). In more than one case, the process of restriction occurs in a rather dogmatic way: we say things

[30] On intergroup dynamics, see at least Tajfel (1981) and Brown (1989).

[31] It was for instance Sebeok himself to dismiss once and for all the idea that only human communication can be of symbolic type.

like, "culture is the integrated pattern of *human* knowledge, belief, and behavior that depends upon the capacity for learning and transmitting knowledge to succeeding generations" (as appearing in the Webster dictionary), but we do not explain why learning capacities and knowledge transmission (extremely common phenomena in most animals) should be exclusively "human": we just take it for granted. As we mentioned already, this is the stage where another problem may arise—what we already called "friendly fire": more often than not such restrictions end up excluding from the picture also several human communities or human subjects, creating discrimination that, *in this historical moment*, is not really acceptable. Some of the most recurrent definitions of "music" are for instance so circumscribed that, while keeping out dangerously-melodic species like humpback whales or nightingales, they also end up discriminating the majority of musical cultures around the globe, plus illustrious musicians like John Cage or Karl-Heinz Stockhausen.[32]

---

[32]It is really emblematic that Dahlhaus and Eggebrecht (1985: 143), in concluding their book *Was ist Musik?*, and after putting so much theoretical effort into trying to circumscribe and define the musical phenomenon, end up admitting that it is pretentious to expect a definition to emerge at the end of the book. According to times, scholars and contexts, indeed, music has been seen as something mathematical, ethical, emotional, referential, nonreferential, symbolic, dynamic, and so forth. Thousands, literally, have been the attempts to define it, and two (and two only) are the points on which a certain agreement among scholars seems to occur: music is an event related to the presence of sounds; and, fatally, music is a typically human phenomenon (or, as Blacking puts it, music consists in "humanly organized sounds").

Now. With the permission of Cage and his silences, I think there is very little to argue about the first point, being music, in physical terms, a sound expression (not only, but at least). And, in fact, the idea itself of using silence as a form of musical expression (as in Cage's famous *4′33″*) confirms, rather than contradicts, the intimate, unbreakable relation between music and sound events. The second point represents, on the contrary, a big theoretical problem, and the unbiased reader may wonder what are the (empirical) bases by which we would be allowed to consider music only a human phenomenon. Let us take, one for all, the definition of music provided by the Encyclopaedia Britannica:

> Music: art concerned with combining vocal or instrumental sounds for beauty of form or emotional expression, usually according to cultural standards of rhythm, melody, and, in most Western music, harmony. Both the simple folk song and the complex electronic composition belong to the same activity, music. Both are humanly engineered; both are conceptual and auditory, and these factors have been present in music of all styles and in all periods of history, Eastern and Western.

Apart from other theoretical implications (not interesting for us, here), this authoritative definition underlines exactly the two points: music as a sound phenomenon, and music as a human phenomenon. Nobody is taking care of explaining, however, why music is just human. What are the specific limits which prevent other sound-producing animals to be in the category of music producers? Nobody is providing explanations on the claimed incapacity of animals to produce music. I do not claim it is a matter of bad faith, it is just that this aspect is conventionally taken for granted, and there is simply no interest in considering the opposite hypothesis. In the whole of contemporary musicology, examples of mere mentions of this argument—apart from explicitly zoomusicological literature—are extremely rare (for a full argumentation of this particular problem, see Martinelli 2009: 102–116).

3.3.2.1 The Semiotic Animal

The main problem remains anyway one: this vital urge, on the part of human beings, to (1) establish their identity on the basis of claimed radical differences with other animals, and (2) setting this separation in form of qualitative dualism, as in human beings on one side and all-other-animals, indistinctively, on the other. The former is basically an anti-Darwinian stand, the latter is anti-Darwinian, anti-Uexküllian, anti-scientific, and—for that matter—anti-semiotic. I discussed at length the rhetorical dynamics of human exclusiveness in terms of dualism in Martinelli (2010: 291–325), so the question will not be further explored here.

Now. There has always been, in one moment or another of semiotic history, (a) "scholarly campaigns" in favor of the uniquely human application of semiotic studies (particularly in the European schools of Saussurean and Greimasian extractions), or (b) attempts to contribute to those processes of progressive restriction of given concepts (semiotics had a go on at least "language", "symbol", "culture", "abstraction" and "signification"), or finally (c) creation (or participation to the creation) of new labels of human distinction (the linguistic animal, for one).

The last, specifically semiotic (and it could not be otherwise, given its name), creation offered, and my case study here, is the concept of the "Semiotic Animal". A colleague and friend like Prof. John Deely, the main promoter of this concept, subtitles it "a post-modern definition of *human being*". As he recalls:

> The earliest record we have of the expression "semiotic animal" in English dates to Deely 1990,[33] in Italian to Petrilli 1998, but both these usages find a considerably earlier ancestor in the test of Hausdorff 1897, written under the nom de plume 'Paul Mongré': [...] "The human being is a semiotic animal" [...] There is, however, this considerable difference in the original German coinage and the later coinages in English and Italian: Hausdorff explains his expression in the terms that would by the mid-twentieth century be associated with the glottocentric minor tradition of semiotics commonly called "semiology". (Deely 2005: 25–26)

By this, Deely means that Hausdorff was an *ante litteram* supporter of the heavily anthropocentric view of semiotics that dominated the scene through the whole 20th century until roughly the 1980s: the linguistic-semiological tradition of Saussure. In this sense, talking about a "semiotic animal" means nothing less than affirming, as Hausdorff did, that the "human nature lies in the fact that, instead of expressing naturally his needs and wants, [man] has developed a language of signs which is conventional, symbolic, and only indirectly understandable" (Deely 2005: 26, translated from Hausdorff 1897: 7).

On the contrary, the semiotic tradition followed by Deely, that red line that connects—among others—Peirce, Morris and Sebeok, assumes the almost opposite stand of "opening" semiotics to the multiple instances of non-human signification, communication and representation, "seeing cultural creation itself as a natural extension of the activities of the semiotic animal according to what is proper to it as

[33]That is, the first edition of Deely (2009), in this book's bibliography.

part of nature" (Deely 2005: 26). In that sense, there is never, in Deely's theories, the denial of the animal nature of human beings, and in fact (at least according to his own words in various conferences and talks), the notion of semiotic animal is apparently conceived for overcoming the human-animal dualism and creating that continuity-discontinuity dialogue envisioned by Prodi (1983: 180):

> The anthropos as semiotic animal is an interpretant of semiosis in nature and culture alike, that can only be because the ideas of this animal, in their function as signs, are not limited to either order, but have rather [...] the universe in its totality [...] as their object. (Deely 2005: 27)

And, more to the point:

> At the heart of the difference between the human Umwelt and the Umwelt of other cognitive organisms is the "idea" in this specifically semiotic sense: the relationship itself that constitutes signification is grasped in its proper being at once imperceptible and distinguishable both from a given signified and from a given sign-vehicle – and therefore as detechable from any given vehicle and attachable to any other vehicle, as well as directable to some other object, or to the same object only, in its new attachment. (Deely 2009: 89)

What does this mean in practice? It means that the very species-specific characteristic of the human being is the capability of emancipating its semiosis from the constrictions of its own Umwelt and (a) access other organisms' Umwelten (also, but not only, at a purely speculative level), and (b) creating possible worlds, i.e., Umwelten that are distant in time, in space, or that are not even existing at all, but are created through language (there is no such a thing like mermaids, yet there is such a thing like mermaids, at the very moment at least one person linguistically produces a concept, or an idea, of them). Such abilities, Deely points out, do *not* make the human being *something else* than an animal, yet they establish the ground for human biological identity. The identity being: the human beings, unlike other species, are *semiotic* animals.

John Deely is one of the firmest supporters of zoosemiotics, and a most credible 'heir' of Thomas Sebeok, of whom he was also an intimate friend. In that sense, the last thing one might suspect is that he is anthropocentrically-biased. The fact that this time the sign of human distinction is (a) limited to a tiny particular in the entire spectrum of possibilities previously explored,[34] and (b) expressly formulated in such a way that it is made unmistakably clear that the human being *is* an animal in all respects, must be certainly regarded as the fairest step taken so far in this special category of speculations.

The present observations indeed are of purely methodological nature, and partly call into question the research ethics of this type of intellectual enterprise. In this respect, they must be read as general observations, which only use Deely's

[34]There is no doubt that being able to create possible worlds is a much milder form of distinction than, e.g., being "social" or "spiritual", or "rational", all being traits filled with countless implications at different levels of cognition and behavior.

formulation as an excuse, and not at all as the main target.[35] The point we are trying to make, in the economy of this entry of the Numanities manifesto, is that arguments, within humanities, are often delivered in speculative form, not only when speculation would be the sole intellectual tool at the scholar's disposal (a situation that would be altogether fair), but also when empirical research is (a) possible, and (b) needed.

With this in mind, there are a number of questions that are left unanswered, ambiguous and possibly biased, in this and other formulations. First and foremost. The motivation. Why is it so important, for a semiotician (and for a scholar in general), to establish a clear sign of qualitative distinction with other animals. The question is not rhetorical, and in fact neither ethical (although it may appear so), but mainly historical and anthropological. There seems to be a need, what we already called a "vital urge", on the part of human beings, to define themselves as an ingroup possessing certain characteristics, in relation to a bigger outgroup which does not possess them.

At this specific historical moment, the other animals are identified as that outgroup, but it was not always like this. It is often forgotten (because it looks absurd with today's eyes) that the original *Scala Naturae* conceived by Aristotle (from which the notion of rational animal is based), is not only an expression of human-animal dualism. The truth is, that was not a dualism at all. It was a comparison between a single category of privileged human beings (those who were also male and free) over three main categories of unprivileged: women, slaves and, yes, other animals. If a dualism must be established in the Aristotelian framework, thus, its definition should rather be borrowed from Peace Studies jargon: it was an oppressor-oppressed dualism, in Dussel's terms, or a center-periphery one, in Galtung's.

In general, the shaping of the ingroup identity was, until very recently, not necessarily characterized by the urge of defining "humanity" *in toto*. It could take all sorts of disguises: male people, free people, white people, western people, sane people, rich people, Christian (or Muslim, or Jewish, etc.) people, heterosexual people, and so forth. What happened (at different times and places of the 20th century) is that most (to a certain extent, all) of these forms of discrimination finally and thankfully came to an end, at least on a generally-accepted ethical level. There is no ground, anymore, for anybody to say that men are rational and women are not, without being taken as a chauvinist pig, or as a cheap joker. Whoever advances any post-Nazi claim that people of Caucasian ethnicity are more intelligent than, say, Afro-Americans, raises immediately general irritation, and it also takes a special effort to consider a similar statement as just a bad joke.

Given this picture—forgive the sarcasm—what is left to those poor intellectuals who still want to exercise their anthropological urge to be part of a small elite of elected individuals? Racism, chauvinism, classism, Eurocentrism, and all the rest

[35] And this is said also as one more demonstration of my esteem for Deely as a scholar, which has nothing to do with the few critical remarks advanced in this text.

are gone. Vanished in that annoying black hole called civilization. How to reconcile the primary need for discrimination with the modern pressure of a civilized and tolerant outfit? The answer is what philosopher Peter Singer named *Speciesism*, the genesis of which he explains exactly in the terms here illustrated:

> In recent years a number of oppressed groups have campaigned vigorously for equality. The classic instance is the Black Liberation movement, which demands an end to the prejudice and discrimination that has made blacks second-class citizens. The immediate appeal of the black liberation movement and its initial, if limited, success made it a model for other oppressed groups to follow. We became familiar with liberation movements for Spanish-Americans, gay people, and a variety of other minorities. When a majority group—women—began their campaign, some thought we had come to the end of the road. Discrimination on the basis of sex, it has been said, is the last universally accepted form of discrimination, practiced without secrecy or pretense even in those liberal circles that have long prided themselves on their freedom from prejudice against racial minorities.
>
> One should always be wary of talking of "the last remaining form of discrimination." If we have learnt anything from the liberation movements, we should have learnt how difficult it is to be aware of latent prejudice in our attitudes to particular groups until this prejudice is forcefully pointed out. (Singer 1989: 148)

Speciesism, *so far*, is the "last remaining form of discrimination". Considering the encouragingly increasing number of animal welfare legislations in different countries that are extending the basic human rights to other Great Apes, that is, to species belonging to our same family *Hominidae*, it is fair to forecast that the next sensation, in terms of discrimination trends, will be called "familyism". And this is not necessarily a joke.

There is still ground for speciesist statements without being politically incorrect. There is no formal program for a "he or she" formula, as applied to non-human animals, and the same expression "non-human animals" is easily replaceable by "animals", achieving the same result, in people's understanding, of referring to 'all other animals except the human one'.

So, first and foremost we need to understand *why* such intellectual enterprises seem so important to the scholarly community. Why in general, but also why in the specific context of scientific communities who have (or should have) fully accepted the Darwinian notion of the evolutionary continuum. It is difficult to grasp the coherence here, and inevitably one is reminded of Richard Dawkins' warnings on the great difficulty of complete emancipation from religious beliefs, even in a secular context.

The theory of evolution is at present—from both the philosophical and the empirical point of view—the most reliable scientific theory for explaining life on this planet. Among the millions of important implications of the Darwinian revolution, there is also a firm warning not to discuss *anymore* the differences across species in terms of *qualitative* distinctions. There is no such point in which an organism radically *stops* being/doing something and *starts* being/doing something else. Everything occurring in life is a natural continuation of a process in continuous development, with traces detectable in different living forms, and with aspects that are only *quantitatively* developed in certain forms rather than others. However, and that is crucial here, *no more* should scholars be allowed to place a threshold at any point of evolution and affirm that "this is it, from now on it is a completely different matter, whose features concern only this species, and not all the others". Also, and this is the lesson instead learned from Uexküll, it is time to quit this *human and-all-other-animals* dualism once and for all. Human beings are a species, not the conceptual counterpart of the Animal Kingdom.

I believe there should be a point in which humanists realize that, if they cannot *prove* those speculations they produce against what has been already proven by science (especially when they make such big claims like "what *really* makes humans special", which is proportionally not unlike the "what really makes Arians special" of a few decades ago), then they should rather focus on something else.

This is certainly one of the sensitive areas within semiotic discussion, something that quite a few semioticians do not seem to grasp entirely, probably because they think that semiotics, with its own tools and procedures, is alone able to overcome certain problems. The Peircean process of endless semiosis is awfully charming, and it is understandable that a few scholars are tempted to turn it into a straight philosophy of science. But then, these scholars should not be surprised if they end up being labeled as rather arrogant, and by consequence the whole of semiotics is regarded suspiciously. This is why, besides a need to support Darwin against fancy retro-futuristic notions (i.e., those notions that, while attempting something new, are again restating pre-Darwinian anthropocentric ideas), another statement is also called for: when semiotics aims to discuss science, it ought to be scientific. Period. If statements like a supposed qualitative diversity of the human being are issued (or, on the other side, the presence of interpretation processes in plants' semiosis), proofs should be provided, not only speculation.

Which kind of leads to a second question: "Have we not learned the lesson yet?". Human intellectual history, as I mentioned, is full of these repeated attempts to create a "special" ingroup, and a specific care, in the last century, was taken to establish this uniqueness to the detriment of other animal species. Now, all these attempts failed. Long ago, or recently. Honorably (like the "symbolic animal" or the "abstract animal", which admittedly took accurate research and speculation to be finally dismissed), or miserably (like the "social animal" or the "cultural animal", which clashed not only against basic scientific inquiries, but also against lay-people and children's observations). Now, what a "reasonable animal" would do, in order not to look like a "pathetic animal", is to behave like a "careful animal", and avoid advancing more hypotheses by using the same old dualistic framework. There is vast ground for building the human identity (again, provided people cannot really help establishing one) on a fair Darwinian level, replacing the digital with the analogical, the concept of differentiation with that of specialization. And, most of all, without making the whole enterprise look like a Freudian phallic drive to show the world that "we do it better" or "we have it bigger".

Plus, apart from scientific arguments, the main reason why it should be time to stop doing this is that it is, simply, a bit childish. Unfortunately and evidently, this is yet not the case. Which leads to the suggestion that the best definition for the human being is that of the "stubborn", or the "frustrated" animal. This animal still seeks for a dualistic, qualitative confrontation, because we humans cannot just accept the idea that we are animals. Or, if we surrender to this scientific evidence, we still need to prove that we are special animals, but not special to the extent that every species is special (like Uexküll's Umwelt theory suggests). No, we want to be special in a special way. Second-degree special. All animals are equal, but some are more equal than others. Orwell did not know how *extensively* he was right.

3.3.2.2 A Few More Bibliographical References, Please?

However, it is not my intention to start an animal rights-inspired monologue, so let us return to the specific discussion, and to the reasons why, I believe, if we plea for more evidence and less metaphysics, and we want to do it as "humanists", we need to deconstruct certain humanistic formulations with care and precision. So, let us go back to the "semiotic animal". Another important question: if we humans are semiotic animals, what type of animals are the other animals? Non-semiotic animals, unsemiotic animals, semiological animals? This does not explicitly appear in Deely's or Petrilli's writings, but it was cleared up during a friendly chat between John Deely and myself in June 2009, on the occasion of the Summer School organized by the International Semiotics Institute in Imatra (Finland). Non-human animals, it appears, are semiosic animals, with an *s*. Therefore, the capability of creating possible worlds is what distinguishes semiotic from semiosic cognition.

So, clearly, a new semiotic threshold is now established and it is set to an extremely high point of cognition: "possible worlds" (after the times, mentioned at the beginning of this chapter, when "intentionality" was the dividing line). Semiotic are those animals that manage to create and access possible worlds, semiosic are those that cannot. Regrettably, it does not seem a great improvement from the old threshold based on intentionality, which achieved the same identical result: humans above, other animals below. Not to mention that at this point it is probably needed to diversify the discipline formerly known as semiotics into two fields: the logocentric (or metalogocentric) "semiotics", and the huge (99 % of semiosis, Sebeok used to say) "semiosics", that is, a field that investigates all semiosis processes, except those that lead to the creation of possible worlds. But, no doubt, Deely for one cannot possibly want this.

Sic stantis rebus, the upper level of the semiotic threshold should be considered as corresponding to the realm of abstract imagination (the one that, indeed, makes possible worlds cognitively possible, if you allow me the play of words). Or should it? Of course, that is not the end of the story: a crucial precondition to this form of imagination is the ability (this also attributed to humans only, in the "semiotic animal" formulation) of using signs and being aware that they are signs. Only with this *metasemiosis*, as Petrilli calls it (1998: 8), the human animal becomes aware that "there are signs", and that these signs can be detached from any association and manipulated to create endless new associations, even imaginary ones. It is metasemiosis that makes mermaids (and unicorns, angels, God, Neverland, Spider Man and Dr. Octopus) possible. That, it is claimed, occurs only among humans.

Now, let this point be the first on the checklist. To tie inevitably metasemiosis with possible worlds, one should assume that metasemiosis *only* produces possible worlds. Because, if it produces something else, then it is not just alien Umwelten that animals are not able to access: they would also fail to access all the other outcomes of metasemiosis, as metasemiosis, too, it is claimed, is a species-specific human phenomenon. There goes the point. The awareness that "there are signs" is a precondition for a million of other cognitive abilities. Deception, for one ("semiotics studies everything that can be used to lie" is a famous quote from Umberto Eco). Then Play. Symbolic representation. Aesthetic semiosis, Ritualization. To a certain extent, also interspecific communication. They all imply (simple or complex) forms of sign manipulation that are possible *only* by being aware that what is being used *is a sign*, and not the actual entity it refers to. And, looking closely, one might even argue that codes themselves cannot be established without an at least vague understanding of the nature of signs.

It is not clear whether Petrilli was aware of the "size" of metasemiosis, and, if so, whether she actually intended that non-human animals are indeed incapable of all this. What is certain is that there is no shadow of doubt, from the present state of *all* animal-related research fields (including zoosemiotics), that non-human animals are unmistakably capable of all these behavioral patterns: there are differences, of course, across species, and it would be very interesting to finally discuss *those*, rather than posing the problem in a dualistic way, but nobody here seems to be interested in that, as the outgroup "animals" is systematically treated as one big cauldron.

Now, having taken care of "metasemiosis" as such, let us return to one of its (many) implications: the possible worlds. Would it be right to assert that no animal except the human one is able to cognitively access alien Umwelten, particularly the imaginary ones? It is a quite interesting question, with several possible answers:

1. One could for instance follow the logics of Peircean abduction, and state immediately that "As the unavoidable precondition for possible worlds cognition is metasemiosis", and "As metasemiosis is empirically existent in many non-human animals", then there are good reasons to think that many non-human animals are able to cognitively construct possible worlds. This reply alone will hardly satisfy the readers, but it is a reasonable point already. Language is not the only condition for metasemiosis: of course it is not. If metasemiosis is equaled with possible world cognition, there is a de facto denial that the latter is produced exclusively by language. And, in fact, there are very good reasons to think so;
2. The concept itself of accessing alien Umwelten may be in fact a contradiction in principle. Taking Uexküll's theories very faithfully, one shall deduce that if we, members of the Umwelt A, manage to cognitively access the Umwelt B, then we cannot really talk about different Umwelten anymore, because the occurrence of this very process would prove that the Umwelt B is simply part of the Umwelt A, therefore not 'alien', and in fact not Umwelt, either. When we talk

about *things*, in a way, they are no longer *things*, but already *objects*, or at least philosophical *objects.*[36]

3. In any case, even accepting the language's capacity to create strong dialectical, philosophical and even rhetorical configurations of alien Umwelten, a few questions remain unanswered. For instance, dialectics, philosophy and rhetorics do not give *full* access to another Umwelt, even though they certainly allow a good deal of reflections and some conclusions about it. This is important to remember, especially at an ethical level: sometimes, indeed, we take too much for granted that we understand of other animals *all* that is to be understood. To make a specific zoosemiotic example, we may understand, speculate and talk about echolocation, but a complete, senso-motorial, psycho-physiological access to a process like echolocation (at least so far) is not possible, and that makes a huge difference (especially when we think of the dominant role that echolocation have in the semiosis of many cetaceans);
4. That is not all. We should not be so enthusiastic about our invention of myths, tales and superheroes, at least not to the extent of founding a new definition of the human being upon it, because all the imaginary entities produced by language are *not* entirely untied to our perception. They might not exist, but they are a combination of (mostly visual) perception-bound elements (a horse plus a horn for a unicorn, a human female plus a fish for a mermaid, a zoo- or anthropomorphic outfit for most gods, etc.).
   We know already from narratology (see at least Genette 1972; Searle 1975; Bruner 1986) that possible worlds are forms of (a) parasitism and (b) limitation of the real empirical world. We know from logic (see at least Lewis 1986; Herrick 1999; Divers 2002) that their ontological status in relation to reality is far from independence, and in particular their modal status completely relies on this relation (the six propositions—true, false, possible, contingent, necessarily true and impossible—are no less than hermeneutic variations applied on empirical reality). We finally know from Freud (1899) and later Jung and the whole of psychoanalysis[37] that the "true" abstract imagination, i.e., dreams (those we cannot even logically reconstruct after waking up because they escape the cerebral centers for thinking activity, which indeed do not function during dreams), belongs to a realm, the unconscious, which is neither linguistic nor para/pseudo/proto/pre-linguistic: it has simply nothing to do with language, but

[36]"Now there is a great difference between an object and a thing, however confusedly the two notions are made to play in popular culture. For while the notion of thing is the notion of what is what it is regardless of whether it be known or not, the notion of object is hardly that. An object, to be an object, requires a relation to a knower, in and through which relation the object as apprehended exists as terminus. A sign warning of "bridge out" may be a lie, but the thing in question, even in such a case, is no less objective than in the case where the sign warns of a true situation". (Deely 2000: 18)

[37]With the possible exception of Lacan, who never gave the impression of exactly knowing what he was doing with psychoanalysis.

with cerebral centers for sensorial and nervous activities. A little too 'biological' for being the "ultimate sign of human distinction".

5. Let us anyway stay on this particular (perception-bound) case of abstract imagination: the possible worlds constructed thanks to the rhetorical-narrative properties of linguistic semiosis (hoping that it is clear that this is an entirely different matter from the truly intangible possible worlds investigated in psychoanalysis). The question is: what elements do we have to be able say that these features of human semiosis are *qualitatively* different from other animals? If many experiments in interspecific communication turned out to be a successful attempt to teach human language to other animals, as they did, in mostly everybody's opinion except semioticians', and if language belongs to the human Umwelt, then it follows automatically that those language-trained animals *did* access the human Umwelt, in fact they accessed what is commonly regarded as the quintessence of it. An extensive mastering of such a powerful device as language, as possessed by humans, certainly allows a drastic *increase* (note: "increase" is a term that does not imply qualitative distinctions) of this type of semiosis, therefore the more refined and articulated our linguistic knowledge, the deeper our cognitive abilities in this respect.
   However, the outcomes of those experiments seriously call into question our exclusivity in this respect, and—at the same time—reinforces the hypothesis that this specific form of metasemiosis is a language-specific feature. Which makes a subtle, yet dramatic, difference. Because these features now mark a difference between language and other communication and modeling systems, *rather than* between humans and other animals. If other animals were able to learn language, at least to some extent, that categorically means that it is in their cognitive potential to learn it.
   And, by the way, it is always surprising to notice that these interspecific communication programs are never considered in their phylogenetic aspect. It has been said that apes learning language reach the cognitive abilities of a young human child. The assessment might be accurate, but, for a change, it is anthropocentrically unfair. The first—very rudimental—traces of a language presumably appeared in the genus *Homo* (4–5 millions of years ago), and took millions of years to take recognizable shape: the anatomical and neurological features of a language-inclined human were developed gradually and slowly, and with a good number of random events. Now, compared to this immense time, it is a bit too much to demand perfection from the first generation of non-human animals who have been introduced to human language (we are talking about less than one century). Alex, Koko, Kanzi and the others have shown that it is in the potential of many species to learn a sign system of that kind: if we were already expecting *Planet of the Apes*-type of scenarios, then clearly we did not consider this aspect. I shall return to interspecific communication later, as it is the other important case study I intend to discuss in this chapter.

6. One point is however missing in the previous argument. The mentioned experiments in interspecific communication provide evidence that other animals are semiotically able to access only alien *existing* Umwelten, not imaginary ones. None of what has been remarked suggests actively that non-human animals are able to invent (let alone conceptualize) an imaginary entity. In this very detail, the notion of "Semiotic animal" may be accurate. No trace of imaginary possible worlds has been found in non-human cognition.

Or, let me rephrase it: no *trace* of imaginary possible worlds *appeared* in non-human cognition. It is a crucial correction, because in order not to *find* something, we first need to look for it. Did we check whether non-human animals have their mermaids and gods? Of course not: how could we? How is it possible to detect a number of signs in a given animal, to separate the apparently meaningful from the apparently meaningless signs, then to analyze the apparently meaningless ones, and finally to understand that some of them are not in fact meaningless but refer to an imaginary entity? Only imaginary entities like Superman can possibly do that.

To my knowledge, the closest we got to a systematic (and, again, empirical!) study of animal imagination are the following cases:

- Robert Mitchell's edited collection *Pretending and Imagination in Animals and Children* (2002), which however tends to present a definition of imagination that is pre-linguistic and closely related to the idea of deception and play.
- Studies like Correia Dickinson et al. (2007), Clayton et al. (2003), Emery and Clayton (2008a), Mulcahy and Call (2006), and Raby et al. (2007), on alternative future-scenarios in the cognition of species like jays and apes. Once again, we might here object that, although 'alternative', those scenarios are always perception-bound, so we do not really solve the "semiotic animal" problem (or perhaps, as it is maintained here, we simply provide another proof that it is *not yet* possible to investigate anything else than this).
- Studies like Shanahan (2006) and Shanahan and Baars (2005), where it is argued (through the adoption of a model of information flow from global workspace theory) that the neurological prerequisites for consciousness, emotion and imagination are typical of the mammalian brain. This model—it has been suggested—can be applied to other animals too (Emery and Clayton 2008b: 135). We are extremely far from semiotics, in these cases, but then again semiotics has not the appropriate tools for investigating these issues on an empirical basis.

Even with those animals that learned human language there is very little we can do. It is known that some of them are often caught performing signs at random, for their own entertainment or in front of their trainers. Since this is a completely speculative environment, why not launch the hypothesis that some of those random signs may actually be the result of their own imagination? Maybe the likes of Washoe, Kanzi and Koko are accessing imaginary Umwelten, and even describing them to us. It may sound like a ridiculous hypothesis, and it probably is, but once we are in the field of speculation, all logically-acceptable hypotheses have a similar ontological status.

To be precise, there are even more authoritative speculations on this subject. Jane Goodall, during her very long and well-known experience with chimpanzees in their habitat, described a group performance that she called a "rain dance" (Goodall 1971: 54). Naturally, as Sebeok (1981: 219) remarks, the meaning of this behavioral pattern is not at all clear. However, what seems to be certain is that chimps dislike rain, thus their dance is possibly not a way—as it is with humans—to invoke water. Is it then to make it stop? It is not known: a display of disappointment is probably the simplest option. But one cannot exclude a certain rough degree of spirituality (i.e., the claim that the very act of rain-dancing has some kind of influence on the weather conditions).

> Darwin, too, was convinced that traces of belief in the supernatural were present in other animals as well:
>
> [M]y dog, a full-grown and very sensible animal, was lying on the lawn during a hot and still day; but at a little distance a slight breeze occasionally moved an open parasol, which would have been wholly disregarded by the dog, had any one stood near it. As it was, every time that the parasol slightly moved, the dog growled fiercely and barked. He must, I think, have reasoned to himself in a rapid and unconscious manner, that movement without any apparent cause indicated the presence of some strange living agent, and no stranger had a right to be on his territory. (Darwin 1871: 67)

And, along with this, Darwin would also push the notion as far as to assert that also religious devotion had a basis in the non-human realm:

> The feeling of religious devotion is a highly complex one, consisting of love, complete submission to an exalted and mysterious superior, a strong sense of dependence, fear, reverence, gratitude, hope for the future, and perhaps other elements. No being could experience so complex an emotion until advanced in his intellectual and moral faculties to at least a moderately high level. Nevertheless we see some distant approach to this state of mind, in the deep love of a dog for his master, associated with complete submission, some fear, and perhaps other feelings. The behaviour of a dog when returning to his master after an absence, and, as I may add, of a monkey to his beloved keeper, is widely different from that towards their fellows. In the latter case the transports of joy appear to be somewhat less, and the sense of equality is shewn in every action. (Darwin 1871: 68)

The hypothesis that only humans have this kind of imagination is constructed on two points chiefly: (a) the fact that humans clearly have this ability; and (b) the fact that no similar ability *appeared* in other animals. Period. This allows for a fair amount of speculation on the subject, but should not reasonably allow us to *affirm* that non-human animals, *therefore*, are not capable of imagination. Otherwise, we witness the same type of argumentation of those people speculating that there are no other intelligent living forms except those inhabiting the planet Earth. The argument is: we have not seen any, so far (or, worse, none has shown up yet), therefore they do not exist. A far cry from providing evidence of the absence of such living forms, especially when we consider how little of outer space we have so far been able to explore.

This latter point creates an interesting parallel with the affirmations on human qualitative distinction within the Animal Kingdom: such affirmations are indeed, in most of the cases, not supported by solid knowledge of animal behavior. When I talk about "empirical evidence", I do not necessarily mean that humanists have to design and perform an actual experiment to prove their theories. It would already be a fantastic achievement if they could at least base their reflections on the works of those scholars who did bother to work within an empirical environment. In other words, speculation for speculation, one prefers to trust Jane Goodall or Charles Darwin, who at least spent a life among non-human animals, so they might have a clue of what they are talking about.

All in all, in fact, this is the thing that puzzles the most in this whole enterprise: the lack of specific competences. Nearly everybody in semiotics sooner or later has a go at non-human animals and on the human qualitative difference with them, yet nearly nobody has had a specific training in animal studies to make them entitled to issue such important statements. With almost the sole exception of Sebeok himself and few others, essays on human uniqueness within the animal kingdom contain a remarkable amount of references related to human knowledge, humanity and human behavior, and at the same time they display an astoundingly poor number of references to animal studies. To speak of what humans do and are, as compared to what other animals do not and are not, apparently requires only a solid knowledge of what humans do and are. No care is required to actually check whether it is *true* that other animals do not or are not.

Defining the semiotic animal (Deely 2005) is a 95 page-long book, which, despite its brevity, still manages to contain no less than 134 bibliographical references. It may read even as excessive, but in fact it is rather typical, considering John Deely's impressive knowledge of philosophy and semiotics. There is, however, an unpleasant surprise: in a book that discusses so extensively the uniqueness of humankind over the rest of the Animal Kingdom, one would expect that, if not half, at least one third of the references would actually come from the enormous catalogue of studies in ethology, zoology, sociobiology, primatology, and so forth. Exactly because, if one says that Kenyan people, *unlike Belgians*, are such and such, it would be normal to expect that a comparative study of *both* Kenya and Belgium was performed, and that a respectable balance in the sources concerning both countries was displayed in the bibliography. Well, apparently, references to Belgium are not necessary, because Deely (2005) contains no more than two (two!) references from animal studies: von Frisch's classic on bee-dance (1967) and a lesser known article on balloon flies (Kessel 1955). Two out of 134. Not exactly a fair contest (leaving aside the fact that the most recent reference dates to 38 years before the publication of Deely's book).

And this is just one example of many. "Bodies, Signs and Values in Global Communication", one of the manifesto-articles in which Augusto Ponzio and Susan Petrilli announce the birth of Semioethics to the world, has a very strong incipit, for which no evidence whatsoever or reference is provided:

> Similarly to nonhuman animals being-communication in human animals presupposes the construction of worldviews on the basis of a species-specific modeling device. However, while modeling and being-communication identify in nonhuman animals, in human beings they do not. The specific modeling device in humans, also called language, allows for interpretations, evaluations and responses. Humans are endowed with a capacity for metasemiosis or "semiotics" which presupposes language which is a species-specific primary modeling device. Syntax, deconstruction and reconstruction, the engenderment of infinite possible worlds, the capacity for "semiotics" thus understood, therefore, for evaluation, the assumption of responsibility, inventiveness, creativity and the capacity for planning are all prerogatives of language. (Ponzio and Petrilli 2008: 113)

In this case, we have a 34 page long article, with 58 references. In an article stating that only humans, unlike other animals, display the abilities of interpretation, evaluation, response, syntax, deconstruction, reconstruction, assumption of responsibility, inventiveness, creativity and planning, the total amount of references about animal behavior and cognition is just zero.[38]

As to the reasons for this peculiar phenomenon only guesses can be made: maybe there is some kind of stereotypical *topos* in human sciences not to treat the animal subject seriously (a topos probably encouraged by the extensive use of animal characters in kids' programs); maybe animals is one of those topics that everybody likes to discuss even when an acceptable background is lacking (probably because they are one of those things with which we are in a constant and multiple relation anyway); maybe—once again—the vital urge of defining human identity makes people rather tolerant towards their own prejudices. Maybe something else. Maybe, in this specific case, semiotics enjoys being an exclusive club where only semioticians and cryptosemioticians (i.e., many philosophers) are allowed the right of speech,[39] while—for instance—zoologists, ethologists, sociobiologists are not. It is a fact, anyway, that the animal subject holds a record for being one of the most trivialized topics in the whole sphere of (at least) humanities.

A last remark, before drawing some conclusions. At this point of the discussion, someone might have the temptation to change the cards on the table. Once the option of the semiotic-narrative possible worlds is removed, why not trying the "actual" possible worlds, that is, those related with Freudian studies on dreams and unconscious? What about dreams? Do animals dream? We can be sure that, in the hands of some human scholar (maybe a semiotician), this topic would quickly become the new "ultimate sign of human distinction": after all, as shown, it seems it is enough to affirm that animals do not dream. Who needs references?

If, however, we care to check some specialists' literature (that is, in this case, neurology and other brain sciences), and particularly some recent ones, we would find that the dreaming activity of non-human animals has been proved, in all

[38]Let alone the fact that in general, in all these works, there is not an even vague, *en passant*, trace of the studies mentioned earlier, like Mitchell (2002), Goodall (1971) or Darwin himself.

[39]This would also explain the fact that in both Deely (2005) and Ponzio-Petrilli (2008) the information regarding non-human semiosis are either produced by the authors themselves, or borrowed second-hand from the interpretations and re-interpretations of Sebeok, who at least took the trouble to check from direct sources.

degrees of complexity (i.e., far beyond the stereotype that cats simply dream of catching mice), and in different research programs. Interesting readings, in this respect, might be Smith (Carlyle) (1995), Siapas and Wilson (1998), Poe et al. (2000) and Louie and Wilson (2001).

3.3.2.3 A Plea for Fairness

Having pointed this out, what is left is a very simple observation, perhaps another plea. There is nothing wrong, nor abnormal, in wishing to define the identity of the category one feels s/he belongs to. The formation of ingroup-outgroup dynamics, we learn from social sciences, are the result of a rather basic need to cope with the complexity of reality (see Allport 1954, for instance). What is heavily biased is the second step of this process, i.e., constructing this identity while actively deconstructing the outgroup's identity (and consequently placing the ingroup in a dominant position). In an ideal "live and let live" type of world, the outgroup's identity is by no means a concern for the ingroup. The identity of the ingroup is not less consistent, if the issue of the outgroup is *not* addressed.

A person may characterize him/herself as many things, let us say: Jean-Paul by name, French by nationality, teacher by profession, single by marital status, rugby player and jazz listener by hobby. A bunch of characteristics of this type are certainly very relevant for this Jean-Paul to define his identity: it is obvious, indeed, that these and other conditions/activities of his make him a distinctive individual, clearly recognizable in many circumstances, and so on. That is, Jean-Paul clearly has an identity. Now, at no point of this process, which led to the definition of somebody's identity, was there a need to outdo one or more persons who are *not* Jean-Paul: our French friend does not need to add to his profile that he is such and such, *unlike* Pedro, William or Mariangela. This step adds nothing to the process. If Jean-Paul takes that step, it is not identity he is after: it is domination, it is discrimination. And if, on top of that, Jean-Paul compares himself to Pedro or Mariangela, without really knowing them, then, after domination and discrimination, one also has to write stereotype and prejudice. All of a sudden, Jean-Paul's love of rugby and jazz becomes very marginal compared to his strong identity trait of being chauvinist, ignorant and arrogant.

If scholars understand this, they might possibly understand also that the entire enterprise of human identity definition does not carry the compulsory requirement of additional comments on other animals. Particularly not if these scholars do not possess the necessary knowledge for making such comments (this way also ending up putting in question significant achievements of modern science, like the Darwinian Revolution). A discussion on human sociality, rationality, culture, language, etc. remains extremely interesting and important, if proposed by sociologists, psychologists, culturologists, semioticians, etc. Let the other animals be taken care of by those who have devoted the necessary amount of time and energy to their study.

3.3.3 Case Study 2B: Language and Interspecific Communication

Another pillar of semiotic research is the question of language and its hypothetical applicability to non-human animals. I have been interested in this topic since Martinelli (2006, 2007), and the question remains very controversial (not only in relation to my own research, of course, but in general). In an interview he gave to Susan Petrilli, back in 1987, Sebeok declared the following:

> It is clear that semiotics is interested in two aspects: the study of verbal semiosis (i.e., linguistics) and the study of non-verbal semiosis. However, what most semioticians who are illiterate in biology fail to grasp is that non-verbal semiotics is an enormously wide field, that includes not only the non-verbal human behaviour – which is about 99 % of what human beings do – but also an entire, vast world of millions of animals. In addition, it includes the semiosis of plants and other forms of semiosis that occur within our body, such as the genetic code, the immunological code and other similar mechanisms. Therefore, in terms of pure quantity, it is non-verbal semiosis that largely prevail on the verbal one. (translated from Sebeok 1998: 23)

Since language belongs to the remaining 1 % of the examples provided by Sebeok, then the general assumption of semioticians is that language is a species-specific human feature. I am not here interested in proving whether this assumption is right or wrong, but rather—once again—in discussing the methodology of a statement which—like the previous case study—is a strong one, one that in an ideal world should be corroborated by empirical evidence, or at least—as we discussed above—solidly informed by existing empirical studies. How did semioticians (who are in fact methodologically well-equipped to discuss language) discuss this delicate issue, and what made them so confident that non-human animal communication is in none of its features classifiable as "language"?

The following observations will serve as points of departure for our Numanities-sensitive analysis of the problem:

1. It is not clear whether language is a human species-specific feature or not: one has at the same time proofs that other animals do not *use* language, and proofs that they are able to *learn* it, at least to a certain extent. Within such a picture, the criteria for species-specificity become more ambiguous;
2. Even if it was, one cannot use it as an excuse for creating, once again (after Aristotle, Descartes, religions, etc.), qualitative differences between the human animal and other animals. This book—let me restate it—is fully supportive of Darwinian theories implemented with Uexküllian ones. In both cases, there is no trace of *qualitative* differentiation between humans and other animals. A species-specific trait develops naturally and analogically in one species, from a (cognitive, evolutionary, adaptive) basis shared with more species, so it still bears traces of that common basis. There are zoosemiotic aspects also in language;

3. What exactly language is, is something that scholars are far from agreeing on. Sebeok himself had the rather provocative opinion (which he held firmly, as no provocation at all) that language is *not* a communication device, but primarily a modeling system. Other fields of inquiry would strongly oppose this. And anyway, a common definition of language is far from having been reached;
4. Among semioticians, however, a rather solid opinion on the definition of language seems to have been established. And, among other things, that opinion appears to be very critical towards those many studies (particularly in primatology and psychology) that seem to have proven that other animals are able to learn human language. The fact that this contrast and the one in point 3 exist are enough to consider "language" a critical topic for humanities, one which therefore deserves a thorough[40] discussion in this book.

3.3.3.1 Main Arguments and Problems

It cannot be a realistic aim, here, that of going through the (many) definitions of language and the descriptions of its features, if not in terms of a very general

[40]With such premises, a question will arise spontaneously in the reader (particularly those who fear the excesses of reductionism in science): is there in humankind something *at all* that is *not* applicable to other animals, if not even language may aspire to be an exclusively human subject? This is at the same time a very easy and a very difficult question.

On the one hand, one still has the obvious and scientifically unavoidable fact that the human being *is* an animal. Nothing more, nothing less, and nothing else. Everything a human does is something that an animal is doing. So, nothing human *completely* escapes a zoosemiotic insight. I am not saying that because humans do x, then *all* animals do x. The correct formulation can be made into a syllogism: *Homo sapiens* perform the x pattern; *Homo sapiens* belongs to the Animal Kingdom; the x pattern is performed within the Animal Kingdom. This statement is logically correct, unlike the following: a sparrow does not perform the x pattern; a sparrow belongs to the Animal Kingdom; the x pattern is not performed within the Animal Kingdom.

On the other hand, discussing topics like language, or art, or culture, does not mean to discuss what animals species-specifically do with them. And that particularly applies to human beings, who have of course developed an endless number of scientific/academic fields (incuding the semiotic ones) that are specialized in analyzing these activities. A zoomusicologist may have a hand (and rightly so) in the definition of music, its origins, the components that makes it a biological phenomenon, the notion of performance, the universality of the dance-music association, etc., but cannot (and would not care to) comment on Busoni's transcriptions of Bach, history and evolution of the French horn, differences and similarities between Miles Davis and Chet Baker's versions of *Summertime*, or the use of Pro-Tools in Radiohead's recordings. This is a job for scholars in "human" music. For the same reason, we may also want to discuss what nightingales, or humpback whales, or wolves *species-specifically* do with their music.

The same, of course, goes with language, regardless of a scholar's intention to prove or disprove its existance (as a general phenomenon) in other animals.

introduction to the subject. This section will instead focus on the entire discussion about those attempts to teach human language to other animals, with the explicit goal of questioning the way semiotic research has tackled the topic, and of seeing if the debate can be reformulated in numanistic terms. We shall start from the realization that crucial aspects of the issue have so far been underrated or even missed. In particular:

- The specifically semiotic achievements of the Interspecific Communication Experiments (ICE, from now on: I am of course talking about those various attempts to teach human language to non-human animals, particularly but not only great apes, via such means as sign language or symbolic tokens);
- The connection between these achievements and the notion of language;
- The consideration of the ICE, as research carried out on individual non-human subjects, with specific background, attitudes, etc., in relation to the results achieved;
- The several implications of the ICE, in terms of methodology, outcomes and, in a general scientific sense, ethics.

Most of the reflections proposed here are—to say the least—controversial, and certainly deserve more thorough scrutiny: as a partial justification, I once again would like to invoke the alibi of the pamphlet-like format of this book. As I said, we are not aiming at confirming or dismissing a given statement, but rather to see if a humanistic investigation did its job properly, and was not affected by some kind of bias.

As one may have guessed already, the semiotic discussion on language, and its supposedly human species-specificity, relies—for the most part—on Sebeok's reflections about language itself, and about the ICE carried out by several psychologists and primatologists in the second half of the 20th century. Sebeok has been the pivotal figure in this topic, and most of his followers' arguments seem to reinforce, but hardly change, his conclusions. We can summarize Sebeok's main points in the following list:

1. *Language is a species-specific human device. The ability of language acquisition is a result of the mental capacities of the human being.* "The word 'language' is sometimes used in common parlance in an inappropriate way to designate a certain nonverbal communicative device. Such a usage may be confusing in this context where, if at all, 'language' should be used only in a technical sense, in application to humans. Metaphorical uses such as 'body language', 'the language of flowers', 'the language of bees', 'ape language', or the like, are to be avoided" (Sebeok 2001: 14);
2. *Before being a communication tool, language must be considered a modeling system, specifically, the, very human, secondary modeling system.* "Language is, by definition, a secondary cohesive modeling system providing humans with the resources for extending primary forms ad infinitum. [...] From a biosemiotic

perspective, the language code can be defined as the cohesive system providing the modeling resources for converting what von Uexküll called 'concrete living existence' into 'active plans'." (Sebeok and Danesi 2000: 108). Identifying language with communication was, according to Sebeok's literary words, "a vulgar error" (1991: 71);

3. *The difference between language and non-human forms of communication (and/or modeling systems) is qualitative.* Quite simply, humans have language, other animals do not, rather than saying that language is a more refined form of communication (or modeling system) as compared to other animals' devices, as a Darwinian approach might suggest. It is precisely its nature as secondary modeling system that makes language a uniquely human feature: "All the animal paleontologists classify generically as *Homo*, and only such, embody, in addition to a primary modeling system [...] a secondary modeling system, equivalent to a natural language. The difference amounts to this: while the Umwelten of other animals model solely a (for each) 'existent world', man can, by means of the secondary system, also model a potentially limitless variety of 'possible worlds'." (Sebeok 1996: 106);
4. *Such a capacity to model possible worlds is expressed* via *three major (and again, exclusive) characteristics of language:* what one may call *distant space-time semiosis*, *narrativeness*, and *linking signs* (these three expressions are not found in Sebeok's writings, but are my own way to summarize his concepts). Distant space-time semiosis refers to the ability to keep track, transmit and reconstruct both recent and remote past events and places, and the ability to articulate projects and expectations regarding both immediate and remote places and future events (in a way not dissimilar to what Hockett called "Displacement"). Such ability is both direct and indirect, the former being related to the personal experience of the subject, and the latter referring to experiences that the subject has not lived or will hardly live personally. In other words, language allows human beings to talk not only about their childhood and about their intention to live in a country-house when they will finally retire from work, but also to discuss the defeat of Napoleon at Waterloo and to wonder about the day when Martians will finally land on Planet Earth.

   Narrativeness refers not only to storytelling (which is still an important feature itself), but mostly the general capacity of accessing and describing alien umwelten, either imaginary or not. Regardless of whether these descriptions are trustworthy or not, what matters here is the fact that in principle, any dialectic account of a given context, environment or reality is *possible* in the human semiosis. The descriptions of the Umwelten of both the species Peking Duck *Anas platyrhynchos*, and of the fictional character Donald Duck (inspired by the very same species) are possible only thanks to the existence of such a device like language.

Finally, linking signs (linking words, in particular, since we are talking about language) refer to a specific aspect within the broader concept of syntax (which, for a change, is not an exclusively human feature[41]), namely those signs that in verbal speech are known as conjunctions, transitions, and prepositions. Linking words are a form of para-signs that do not refer to any other existing entity apart from themselves, and whose function is to create meaningful relations among signs that, by contrast, stand for something else than only themselves. In other words, linking words are yet another confirmation of the capacity of language to create possible worlds, even when those 'worlds', as in this case, are simply signs that are untied from tangible entities.

Most semioticians agree that any other claims concerning the uniqueness of language can be dismissed, once and for all, as anthropocentrically-biased semiological, rather than semiotic, blunders. Those include the use of symbolic signs, the ability to consciously deceive the receiver of a message, the ability (and the tendency) to put into action semiosis of aesthetic type, the so-called metasemiosis, and so forth (more details, from the zoosemiotic point of view, can be found in Sebeok 1990: 77–98, 1981: 210–259, various parts in Martinelli 2010 and Cimatti 1998).

5. *The ICE,* carried out throughout the last decades, *were either a failure or, when apparently successful, fatally biased by misinterpretations of the Clever Hans Effect* (CHE, from now on) *type.* However controversial such a position may be (the experiments conducted were of the most diverse types, and one might suggest that it is dangerous to draw one single conclusion from them), there seem to be an almost absolute consensus on this point, especially among

[41]In general, after Marler (1977), scholars distinguish between *Phonological Syntax* (that is, a combination of sounds that, taken alone, do not necessarily have a specific meaning, or at least not a qualitatively different one from the combination), and *Lexical Syntax* (that is, a combination of sounds that are also meaningful as single units, and that also mean 'something else'). No particular effort is required to find examples of phonological syntax in non-human animals. For instance, in Robinson (1984) and in Mitani and Marler (1989) the phenomenon was studied respectively in Wedge-capped Capuchins and in male gibbons. A more difficult task is to find evidence for lexical syntax, but we have a remarkable example in the alrma sounds of Greater Spot-nosed Monkeys. These sounds are of two types (corresponding to their two main predators): something like 'pyow', which designates a leopard, and something like 'hack', which stands for "eagle". The use of these sounds pragmatically proceeds in the same vein as with vervet monkeys. Recent research, however, has revealed an interesting phenomenon with regard to the Greater Spot-nosed. (Arnold and Zubermühler 2008). They form sequences with these sounds (sort of "pyow pyow hack hack"), sequences whose meaning is not simply "A leopard and an eagle" (or two leopards and two eagles), but means something entirely different, that is, "Let us move to another place". In this case we can talk of lexical syntax, as existing meaningful units are combined in order to produce a new meaning, that is not simply the sum of its constituents. Considering that typical process that often turn iconic and indexical forms of signification into symbolic conventions (like onomatopoeias in human language, which are symbolic signs which however depart from an iconic representation of certain sounds), one might guess that this new meaning developed from an indexical formulation of the type "eagles everywhere, leopards everywhere", which syllogistically turned into a kind of "as dangers are everywhere, this is definitely not a safe place", and refined into the more practical "let's go somewhere else".

Sebeok's followers who do not hesitate to label those experiments as anti- or pseudo-scientific. Martinelli (2010) and Cimatti (1998) seem to be the only two exceptions to this rule.

What I will argue here (drawing indeed from my previous work in Martinelli 2010) is that these five pillars, built upon our (i.e., human) concept of language, present elements of inaccuracy and incompleteness, needing additions in some cases, and revisions in others. The following theses will be defended:

- *Much more than "characteristic", language should rather be considered a species-specific human "need".* The ICE have not only demonstrated the cognitive and practical abilities to acquire language on the part of other species, but also that the human being appears to be the most motivated species in using and developing this tool. Moreover, even though not specifically denied by semioticians, *the constitution of the human vocal apparatus in the acquisition of language plays a crucial role*, possibly more crucial than the human brain (that is, it was the possibility of talking that favored the natural selection of language-inclined brains, not vice versa);
- Certainly, there are very good arguments for maintaining that language is more a modeling system than a communication device. However, *the phylogenetic and ontogenetic bases for language to be acquired remain those of the communication system.* In other words, language proves to be particularly efficient as a modeling system, and other sign systems prove to be more efficient than language as communicative forms. *But*, it should not be forgotten that (a) language originated as a communication system, and (b) communication is the primary reason why language is passed on from parents to offspring (that is, parents teach language to their children for communicating, first of all: exploring and categorizing reality are merely consequences of this process).
- Still, if language is—also or mostly—a modeling system, then, in evaluating the ICE, *what shall be analyzed more thoroughly is if and how non-human animals use language to map their Umwelt, and—possibly—to access the human Umwelt.* It is curious that it is precisely semioticians, who are the primary promoters of the notion of language as a modeling system, that discuss the ICE almost exclusively in terms of "communication" experiments;
- *Linking signs seem to be the only specific characteristic of language that is mastered by human beings only. Distant space-time semiosis and narrativeness, on the other hand, are not exclusive of humans, although reinforced and improved by the employment of language.* More specifically, linking signs are the very upgrade (therefore, once again, the terminology employed here does *not* imply qualitative differences[42]), added to a natural communication system,

[42]Apologies for insisting on this point, but it really is the core of the matter, and there should be no misunderstanding whatsoever, in any of the topics discussed here. It seems this is the lesson semiotics is most resistant to learn.

that allows an impressive enrichment of the narrative and distant-space-time semiosic potential of an individual.

- *The discussion conducted by semioticians on the ICE is at times generic and approximate, and requires a certain* (numanistic?) *revision.* Topics that need to be re-visited are at least the following: evaluation of the methodologies and results of the ICE themselves; distinction within different ICE; the CHE and the entire discussion on anthropomorphism; and the consideration of otherwise ignored aspects and outcomes of the ICE.

Most of the following discussion shall focus on the last point of this list, for several conclusions (or so it seems) about language, and its supposed human species-specificity, did actually depart from this very argument.

3.3.3.2 The Experimental Programs in Interspecific Communication

The history of the ICE is pretty long and articulate: it involved various animal species (mostly Great Apes, like chimpanzees, gorillas, orangutans and bonobos, but also cetaceans and birds), several different tools and methodologies (first spoken language, then sign language and material symbols); it departed from the most diverse working hypotheses and ended with the most diverse results (some positive and some negative); and, scientifically speaking, the results were quite often influenced by the time in which the experiments were performed, going hand in hand with contemporary theoretical trends in psychology, ethology and other disciplines (for instance the programs were very behavioristic in the beginning, and very cognitivistic eventually).[43]

[43]It is difficult to wrap up this long history in few sentences and in a footnote. The first experiments were affected by the (eventually proven) wrong assumption that the vocal and acoustic channel was the only reliable source for a communication system to be considered true language. The idea probably originated from writings by Samuel Pepys (The diary of Samuel Pepys, 1661), Johann Conrad Amman (Surdus loquens, 1692 and Dissertatio de loquela, 1700) and Julien Offray de La Mattrie (L'homme machine, 1748), who all shared the opinion that monkeys could easily be taught to speak. Among the first experiments of this type, the chimpanzees Peter, raised by clinical psychologist Lightner Witmer around 1909, and Joni, raised by ante litteram cognitive primatologist Nadia Kohts between 1913 and 1916. Later attempts to teach spoken language to chimpanzees were performed on Vicki, raised by Keith and Catherine Hayes between 1952 and 1958, and Gua, raised by Winthrop and Luella Kellog during the late 1960s. All the chimpanzees—after years of intensive training—learned a maximum amount of 3–4 words, of the likes of *mama*, *papa*, *cup* and *up*. All of them, however, proved to understand dozens of words. Similar attempts were also made with other primate species, like the orangutans trained by William Furness around 1916, and the gorilla Toto (Maria Hoyt, in 1942). The real problem—as demonstrated in Liebermann (1968, 1972)—lay in the neurological and anatomical incapacity of non-human bodies to articulate sounds in the way humans do. The difficulties of these apes were expressive, not cognitive. Later, in 1977, Irene Pepperberg applied the idea to a species that does not have these expressive limitations, the parrot Alex. This experiment was more successful, and Alex proved to be able to understand and pronounce about 100 English words.

Given such a complex scenario, it is a bit surprising that the account given by semioticians of these experiments has most of the time been generic, reductive, and—from time to time—simply inexact. Most of all, what surprises is how generically the ICE are treated. Very seldom are they taken case by case (and subject

(Footnote 43 continued)

As it became clear that not only verbal communication had to be considered language, the interest in the ICE increased, also because of the growing fame of behaviorism. In 1966, Allen and Beatrix Gardner started a program based on American Sign Language (i.e., gestures, not spoken words) with the chimpanzee Washoe, with remarkable results. Other successful ASL-based programs started in 1972 (by Francine Patterson with a gorilla named Koko), and in 1978 (by anthropologist Lyn Miles, with an orangutan named Chantek). ASL was also used in Herbert Terrace's Nim Chimpsky program. After a few years of apparent success, Terrace had a very famous second thought: he realized that the whole training was heavily affected by the Clever Hans Effect, and—with rare professional honesty—he admitted it.

ASL was not the sole alternative to speech. David Premack, whose project started in 1966, trained the chimpanzee Sarah by using colored plastic symbols that had no iconic or indexical relation with the object represented (e.g. an apple was represented by a blue triangle). The project achieved mixed results: apart from concrete objects, whose acquisition was fast and efficient, Sarah showed ability to learn abstract concepts like "name of" (used as "what's the name of x?" when she did not know how to call a certain object), colors, big-small, equal-different, square-round, etc. She could also answer questions like "what is the color of…". However, suspicions that her training was again affected by the Clever Hans Effect remained high.

Scholars then elaborated the ICE that excluded, as much as possible, any interaction between trainer and trainee. Duane Rumbaugh, in the mid 1970s, attempted to teach language by employing symbols. The trainee was a chimpanzee named Lana. Rumbaugh felt that computer automation could prevent anyone from cueing the animal. The machine was programmed to perform certain tasks, like dispensing food or displaying an image in response to pressing the proper symbol on the keyboard. Lana was then given a special keyboard which had symbols (called lexigrams) in place of regular keys. The trainer was connected to Lana with another, equally equipped, computer. The interaction would thus occur only through the respective monitors. Results were again controversial: Lana learned to formulate requests and to engage in relatively complex conversations, but she would also do so only in response to a specific need (usually, hunger or thirst). Rumbaugh was followed by his wife, Sue Savage-Rumbaugh, who carried out one of the most interesting and successful ICE programs, with the bonobo Kanzi, and with a technique that was still based on the use of lexigrams.

Another ICE worth mentioning, among many that were and are performed, is the program conducted by Lou Herman on two dolphins, Phoenix and Akeakamai, in the 1980s. The training consisted of two simple sign systems: Phoenix learned an acoustic language generated by an underwater speaker, and Akeakamai an adapted version of ASL. In both cases, the words produced consisted mostly of concrete entities, actions, modifiers (like "right", "left", "bottom", "surface") plus metalinguistic signs, such as "yes", "no" and "erase" (used as a signal to interrupt an activity). In order to avoid the Clever Hans Effect, the trainers wore a mask that prevented them from showing facial expressions. Moreover, one trainer would give the message, and another one, without knowing the message, would record the dolphins' answers. The syntax of the messages was organized in sequences, like "surface frisbee bring-to basket" (i.e., on the surface there is a frisbee: bring it into the basket). When a task could not be performed, the dolphins would touch a panel correspondent to "no"; when it was, then, after performing it, they would touch a panel correspondent to "yes". This project, too, achieved interesting results. Phoenix and Akeakamai also proved to be able to understand messages that referred to a distant time. They would memorize the instructions, and—as the conditions became favorable—perform the task.

by subject, a point which is not irrelevant): what we are normally offered is one single pot with all the experiments in it, and a set of conclusions applied to all of them indistinctly. Sebeok almost seems the only one to have acquired his knowledge from direct first-hand sources. For quite a few of his followers, it appears that all they learned they learned it from him.[44] What are these conclusions? Trask (1995: 21–22) summarized them in four basic points:

1. Much of the evidence in the ICE is purely anecdotal: "…it consisted of reports that some particular animal on some particular occasion had been observed to do something-or-other pretty damned impressive" (ibid.: 21). Anecdotes, Trask points out, are not of scientific relevance;
2. The criteria for evaluating the IC tests were too broad to be reliable. "For example, if a signing chimp was shown an apple and asked (in American Sign Language) 'What is this?', the experimenters frequently counted as a correct response any sequence of signs including the sign for 'apple', which is a far cry from the sort of response usually heard from a human child learning a first language" (ibid.: 21);
3. In many cases, the claimed ability of certain non-human animals to actually perform the linguistic task assigned to him/her was reported by the experimenters only, so one has "nothing more than the experimenters' own word for it that the apes were making any signs at all" (ibid.: 21);
4. The most important reason why the ICE should all be considered misleading and anti-scientific is their incapacity to avoid the CHE and—more generally—anthropomorphic interpretations of the results of the experiments: "the critics discovered that the experimental procedures typically used to test the animals were so slipshod that an animal under test could often see its human handler unconsciously forming the required response with her or his own hands, so that it could see what to do" (ibid.: 21–22).

To Sebeok, the "pervasive, insidious penetration of Clever Hans" (1990: 68) in the ICE is the main problem, and he reminded us of it in practically every essay he wrote on the subject. Perhaps, thus, the issue of anthropomorphism deserves a deeper discussion, which is what will be provided later on. First, however, the first three points of Trask's list should be noted.

The first point reveals a certain lack of attention in reading the actual texts reporting the ICE. More careful scrutiny would have suggested that most of the anecdotes are not episodic at all, both within the same experiments and, more importantly, across different experiments. For instance, the use of signs in a creative manner to name unknown objects starting from similar known ones is an 'anecdote'

[44]A most remarkable exception to this rule is represented by Felice Cimatti, who indeed should not be considered a Sebeokian or a Peircean (his theoretical bases deriving more from the likes of Jakobson, Gozzano and others), and whose book *Mente e linguaggio negli animali* (1998) is a rather distinct case in the entire semiotic panorama. A short look at his list of references easily proves that Cimatti did his homework on Savage-Rumbaugh, Premack, Pepperberg & co. very well and accurately.

reported in several cases: Koko used "white tiger" for a zebra, "finger bracelet" for a ring, "elephant baby" for a Pinocchio doll, or "eye hat" for a mask (see Patterson and Linden 1981); Lana asked for an "orange-colored apple", not knowing the lexigram for "orange" (von Glasersfeld 1978: 732); Washoe used "drink fruit" for a watermelon, or "water bird" for a swan (Hill 1980: 336). Then again, it is true that Koko called a zebra "white tiger" only once, or Lana an orange "orange-colored apple". But that is because they were soon taught the signs for "zebra", and "orange", so why bothering to still call those objects in their own temporary way? What counts is not the redundancy of a specific example, but rather the redundancy of a rule (or, more precisely, a modeling strategy): when Koko sees a zebra, but does not know the ASL sign for it, she starts reflecting upon it (metaphorically: "Well, it's clearly an animal, it runs… it has stripes like a tiger, but it's white…"). Quite exactly, Koko models her experience through the language. It must also be mentioned that other ICE subjects, like Sarah, were trained to use signs (plastic symbols, in her case) for "Name of", through which they could ask how to call a certain object. In this way, Sarah did not need to create white tigers or water birds, but she would simply ask directly for the right plastic symbol representing the unknown object.

As for the second point of Trask's list, here is a typical example of the situation that will be explained more thoroughly in the next paragraphs: in this case the fear of making anthropomorphic mistakes becomes an even greater mistake. Everyone who is learning a new language (or any sign system) goes through a transitional period when confusion is more regular than precision. Terms, syntax, pronunciation, not to mention the infamous "false friends", are all problems whose solution may take a significant amount of time. The situation described by Trask of the ICE subjects randomly performing signs until they catch the right one is quite normal, as normal as the criterion applied by the researcher in evaluating these attempts: it is exactly what occurs among humans, with children or with foreign language students. We encourage the right answer, as it arrives, with a "yes" or a smile, and we are keen to accept a few wrong attempts before the right word is pronounced when the trainee is just at the beginning of his/her learning process.

Trask's third point is the most problematic and biased. So, we cannot really trust Premack, Gardner, Patterson and company, since they were the only ones observing their trainees performing the experiments, and they may be lying. Never mind that a lot of footage documenting the experiments is available (a lot is on-line, too); and never mind that the projects were never carried out by one single researcher, but by entire teams, whose members would partly or totally change throughout the years (to mention one, Washoe's project has been going on since 1966, and still continues after her recent death, as in the meantime more chimpanzees have been added to the community), and young assistants, whose names would otherwise remain obscure, could easily gain newspaper headlines with scoop interviews of the type "My boss is bluffing about the apes". What is suggested here is that, since we cannot check, there is a reason to think that the ICE are not reliable. If that is the principle, however, nearly any empirical finding, in any scientific field, is liable to suspicion, with the sole exception of those performed in public arenas or on reality TV shows.

3.3.3.3 The Original Sin of Anthropomorphism

Every discipline has its obsessions. A psychologist knows perfectly that if the article s/he submits to a given journal is not invariably structured with the sequence method-results-discussion, it will not be taken seriously.

The obsession for anybody involved in studies of animal behavior is to avoid anthropomorphism: to interpret a given observation of a given behavioral pattern performed by a non-human animal in a way that looks "too human" is like the original sin. If this happens, the scholar will, slowly but inevitably, slip into the realm of researchers who are seen as näive and not scientifically credible.

The ancient Greek *morphé* means "form", "shape" and "appearance". According to Webster's Dictionary, anthropomorphism can be defined in two ways: (1) the representation of the Deity, or of a polytheistic deity, under a human form, or with human attributes and affections; and (2) the ascription of human characteristics to things not human. The dictionary of ethology, edited by one of the most authoritative Italian ethologists, Prof. Danilo Mainardi, one of the leading Italian ethologists, defines anthropomorphism in the following way:

> [Anthropomorphism] is the tendency to interpret animal behaviors by ascribing to them typically human motivations and goals. Occasionally, this led to the evaluation of animals' behavior in moral terms—and to persecuting them.[45] This mistaken habit of seeing animals in human terms has long affected the interpretation of their behavior, and the most evident examples are medieval bestiaries. Research in child behavior has demonstrated that the tendency to anthropomorphize animals arises spontaneously, and that it disappears only in adulthood, thanks to appropriate education. The same tendency is detectable in primitive civilizations and plays an often relevant role in the relation between humans and pets. Dangerous anthropomorphic interpretations have been, and still are, present in many cases of zoological exposition. (translated from Mainardi 1992: 48)

On a very general level, it is evident that anthropomorphism is perceived by the scientific community as a serious mistake, unable to make any useful contribution to ethological knowledge. As emphasized by Regan (1983: 6–7), if other animals are described as having characteristics that are also human, but not only human, such as "being alive", such attribution does not constitute an anthropomorphic mistake. The problem concerns more critical characteristics. Regan mentions the example of consciousness:

> Now, if consciousness is a characteristic of humans only, then we are guilty of anthropomorphism if we regard animals as conscious; we make more of animals than what they are; we erroneously picture them as humanlike. Suppose the charge of anthropomorphism is made against those who view animals as conscious. How can it be met? Surely not by repeating the attribution, no matter how many times it is repeated, and no matter how many different people repeat it. All repeating the attribution could show is how many people view animals as conscious, and this fact, while of interest in some contexts, is impotent in the present one; however many people regard animals as conscious, it is quite possible that the view held by these people is anthropomorphic. Given the charge of anthropomorphism, and given the deficiency of trying to meet it by insisting oneself, or getting others to insist, that

[45]See the very interesting Evans (1906).

animals are conscious, it should be clear that another way must be found if this charge is to be met. (Regan 1983: 7)

The real milestone for the enemies of anthropomorphism is the notorious Clever Hans Phenomenon. Hans was a horse, owned by a certain Wilhelm von Osten during the first half of the twentieth century. He came to be quite well-known for his purported mathematical computing skills. When asked to solve any kind of arithmetic task, from simple addition up to calculating the cube root of numbers like 103,823 (Cimatti 1998: 146),[46] Hans would amaze everyone by striking his hoof against the ground as many times as the exact result was. His fame grew greater as ever more difficult tasks were asked of him and different verification tests were set (e.g., the tasks were written on a blackboard instead of being vocally pronounced). This went on until a scholar named Oskar Pfungst realized that when Hans was interrogated by someone who did not know the right answers, his hoof strikes were wrong. Working from this observation (repeated and confirmed several times), Pfungst realized that Hans's arithmetic competence was in fact quite low, but that his psychological skills were superior to those of most humans.

Clever Hans, indeed, did not have the slightest idea of how to solve arithmetic problems, but was remarkably skillful at detecting the exact moment when the interrogator was pleased with his response (i.e., the right number of hoof strikes). Unintentionally, the interrogator would send some kind of sign (facial or postural, mainly) which displayed surprise in seeing that Hans had actually reached the right number of hoof strikes.[47] At that very moment, Hans would stop striking the ground with his hoof.

Unfortunately, instead of *also* pointing out how exceptional such psychological skills were, scholars only use this episode to recall how dangerous and misleading anthropomorphism can be (in this case, attributing mathematical skills to Hans). From then on, in experimental contexts, one speaks of the Clever Hans Effect every time the animal ends up being the manipulator, rather than the subject, of the experiment, and—more generally (and more specifically, as far as zoosemiotics is concerned)—when the interpretation of a given animal behavior is biased by the illusion that the animal in question has done something a bit too extraordinary.

John Andrew Fisher's "The Myth of Anthropomorphism" (1990) represents a very efficient attempt to classify the several nuances of the concept of anthropomorphism. Fisher is aware of two crucial aspects: (1) there is extreme confusion in the use and definition of the term; and (2) however one may define it, every scholar takes great care to avoid being affected by it, as one would normally do when dealing with diseases:

> Anthropomorphism is usually regarded as an embarrassment to be avoided. Philosophers and scientists often approach anthropomorphism as an obstacle to be overcome by those who wish to attribute cognitive or emotional states to nonhuman animals. Thus Donald Davidson

[46]Which is 47, for the record, but I, unlike Hans, had to use a calculator to figure it out.

[47]We may imagine someone opening his/her eyes and mouth in amazement, or something of that sort.

> suggests that "Attributions of intentions and beliefs to animals smack of anthropomorphism". Even those who favor animal rights try to avoid being accused of it. Annelle Baler, for example, feels obliged to say, "I see nothing at all anthropomorphic or in any other way absurd in saying that one may break faith with an animal, exploit its trust, disappoint expectations one has encouraged it to have". And Mary Midgley asserts: "There is nothing anthropomorphic in speaking of the motivation of animals". Contrary to this loose consensus, I will argue that there is a considerable amount of confusion about anthropomorphism. I will argue that the mistake or fallacy of anthropomorphism is neither well-defined nor clearly fallacious. There are many different conceptions of anthropomorphism and the common ones do not support their common rhetorical use. (Fisher 1990: 96)

A first separation is made between *interpretive anthropomorphism* and *imaginative anthropomorphism*. Interpretive anthropomorphism refers to "all of the usual cases of ascribing mentalistic predicates to animals on the basis of their behavior" (Fisher 1990: 100). Typical in this case is the description of certain behavioral patterns as intentional actions. Imaginative anthropomorphism is in turn defined as "the productive activity of representing imaginary or fictional animals as similar to us" (Fisher 1990: 100). Such is the case with the whole tradition of fables, myths, cartoons, and fiction in general.

Interpretive anthropomorphism is in turn divided into *categorical* or *situational*. The first case (Fisher 1990: 101) consists of "ascribing mentalistic predicates to creatures to which the predicates don't ever in fact apply". The problem is precisely categorical, for given behavioral patterns cannot be attributed to given species without falling into the anthropomorphic mistake. In the case of situational anthropomorphism (Fisher 1990: 101), an animal's behavior is interpreted "in ways that could possibly apply to that animal in other circumstances, but which do not in the situation in question". It is thus anthropomorphic to ascribe a given pattern to species X in situation Y_1, but would not be in the situation Y_2.

Lastly, Fisher divides categorical anthropomorphism into *anthropomorphism by species* and *anthropomorphism by predicate*. The former (Fisher 1990: 101) consists of situations when "application of mentalistic predicates could be counted as anthropomorphism depending on the species. What wouldn't be anthropomorphism concerning a chimp might be concerning a worm". The latter (Fisher 1990: 101) describes situations when "application of mentalistic predicates could be counted as anthropomorphism depending on the predicate. I have in mind applying the wrong types of predicate".

Then, how dangerous is anthropomorphism for scholars? Quite dangerous, it seems, if some scholars even went as far as to formulate theorems to prevent anthropomorphic (or similar) attitudes. The so-called *Morgan's canon* (named after the English psychologist Conwy Lloyd Morgan, whose life bridged the nineteenth and twentieth centuries) is one of the most classical strategies. It says that "in no case should actions or behaviors be interpreted as the result of a superior psychic faculty, when it is possible to interpret them as a result of an inferior faculty". I shall later on comment on this statement. Along with Morgan's canon, another method, quite typical of ICE programs, is the creation of the so-called *Emotion-Free* experimental context. Every possible unwanted input from the experimenters that might result in a CHE (facial expressions that might unconsciously suggest to the subject of the ICE the solution for a specific task, affective—therefore, not anymore scientific—relation established with the subject of the ICE, etc.) is made sure to be avoided. Researchers, then, wear masks, observe the experiment unseen from another room, leave the subject of the ICE alone in the laboratory, avoiding any kind of interaction, and so forth. Not unpredictably, the completely Emotion-Free ICE built up throughout the last decades ended up in a failure, which, according to many semioticians, is the ultimate proof that: (a) non-human animals are not able to learn language; and (b) the CHE is the real core of the problem. When avoided, the ICE reveal their true nature, i.e., misleading non-scientific practices.

3.3.3.4 How Semiotics Criticized the Experiments on Interspecific Communication

Sebeok's criticism of the ICE was not limited to the CHE, and—by all means—was articulated in a much deeper way than Frisk's summary (which was intended as a summary of a number of positions on the topic). In 1980, a collection of essays entitled *Speaking of Apes*, edited by Thomas and Jean Sebeok, was published and soon became the major point of reference for semioticians on the topic of the ICE. It is in the extensive sixty-page long introduction (Sebeok and Umiker-Sebeok 1980: 1–60) to this book that one finds the most detailed illustration of Sebeok's views on the ICE. In the text, the Sebeoks go through nearly the entire panorama of ICE, presenting their doubts or occasionally their resolute rejection. Their arguments are in all cases well-formulated and convincing, and it is probably this characteristic (together with the huge—and deserved—respect towards Sebeok's work in general) that made the approach of Sebeok´s followers on the topic as confident as previously described. Semioticians' following observations on the ICE proved to be rely entirely upon this very text, entitled "Questioning Apes", showing little, if any, advancement from there, in a time (the 1980s and then the 1990s, up to nowadays) when ICE programs were growing rapidly and—in many cases—successfully.

Thirty-five years after, it thus seems reasonable to take a slightly more proactive attitude towards that text, in the light of what has happened in the meantime (let us here recall the importance of the birth—in 1976—and the definitive establishment —in the 1990s—of the cognitive branch of ethology, within animal studies), but also as a consequence of a more critical reading. Maybe some things, although still well-formulated, are not as convincing as they seems.

To start with general remarks, "Questioning Apes" may appear not to respect in full the non-written etiquette of scientific compilations. *Speaking of Apes* is a collection of essays that intends to gather the main positions against *and* in favor of the ICE. The editors of such a work, in writing the introduction to the book, should certainly summarize and contextualize all these positions, but, at the same time, they should not to take too strong a stand in one direction or another. Or, if they do, that should be made by referring to external sources, in order not to discredit the book itself, as a collection made of good essays for one half, and rubbish for another. In other words, the editors should not exploit the chance of writing their introduction *after* all the other texts have been delivered to them.

Regrettably, "Questioning Apes" takes all possible advantages to make it clear how the ICE programs are here regarded as anti-scientific and fallacious, and how naïve the pro-ICE essays within the same book are. Thus, by, say, page 20 there have already been attempts to prejudice the reader against an essay that appears on page 200.

The only ICE scholar spared Sebeoks' criticisms, and in fact highly praised, is Herbert Terrace, who conducted an ASL-based ICE program on the chimpanzee Nim Chimpsky (a witty pun on Noam Chomsky, of course). The program ended up being a complete CHE-affected failure, and Terrace—with noble self-criticism—admitted it with no excuses, thus becoming a hero of the anti-ICE supporters:

> The honesty of Terrace and his colleagues in reporting this and other procedural steps which others, we suspect, omit from their accounts, is to be applauded. It makes it possible for the reader to assess more accurately the reliability and import of the data presented. (Sebeok and Umiker-Sebeok 1980: 13)

The procedures that the Sebeoks are here referring to do indeed reflect the serious bias of the ASL trainers on Terrace's program: they tended to revise their reports after discussing among themselves, thus, in the end, they agreed on many more points than in their previous observations of Nim's behavior. Indeed, comparing each other's observations may be useful for correcting (rather than provoking) individual mistakes, but clearly the chance of messing up the data collected is higher than the chance of improving them. However, the additional step taken by the Sebeoks in commenting upon Terrace's case, is to use it as a synecdoche for all the other ICE programs. The history of the ICE reports both unsuccessful and successful programs, the latter being more recent and numerous. Terrace was simply one who made some mistakes and admitted them. That makes him a highly-respectable and honest scholar, but not the proof that ICE programs are all biased and deceiving.

A last general remark, before getting to more specific considerations, is the fact that the Sebeoks seem not to have been extremely interested in visiting the sites of those research programs. The only instance they mention is a 1969 visit to the Gardners' house in Reno, Nevada, where the Washoe-project was taking place at the time. The description of that visit does not seem too far from how Soviet authorities would selectively guide delegations of western communist parties through the (few) decorous areas of USSR:

> During our stay in Reno, in 1969 [...] we were, as guests, necessarily at the mercy of our gracious hosts, the Gardners, in terms of where and when we could view Washoe, and what additional experiences of the project – in our case, heavily edited films of the chimpanzee, casual testing of her signs in our presence using a box of well-worn objects, and discussions with some of the animal's trainers – we were allowed to have. (Sebeok and Umiker-Sebeok 1980: 29)

This version of the events may be counter-argued in at least six ways:

1. In 1969 the Washoe-project was at the very beginning (it started in 1966 and, despite Washoe's recent death, it is still going on, with many other chimps—including Washoe's offspring—being introduced into the project), and it was only during that year that the first report of the project was published (Gardner and Gardner 1969);
2. A research laboratory, even if located in a private house (as in Gardners' case), has its rules and limitations, and people cannot simply do and check what they want just because they do not trust their hosts;
3. The "heavy editing" of the films was simply meant to spare the spectators hours and hours of totally uninteresting material;
4. A chimp is after all still a chimp, i.e., a wild animal, and it is not always predictable how s/he would react to the presence of a stranger.[48]
5. This is a scientific experiment, not a circus show, and the chimpanzees are not trained to *perform* before any audience, but they rather need to trust their interlocutors before actually interacting with them; finally
6. What makes the reader a bit uncomfortable is how energetically the Sebeoks express their reservations not only about the experiments themselves, but also about the honesty of the IC researchers.

More importantly, there is no information about any other visit to any other IC project, although invitations were not lacking. In a 1998 interview for the New York Times, Sue Savage-Rumbaugh, responsible for the Kanzi project, when posed the question "Many in the scientific community accuse you of over-interpreting what your apes do" answered as follows:

> There are SOME who say that. But none of them have been willing to come spend some time here. I've tried to invite critics down here. None have taken me up on it. I've invited Tom Sebeok (of Indiana University) personally and he never responded. I think his attitude was something to the effect that, 'It's so clear that what is happening is either cued, or in some way over-interpreted, that a visit is not necessary.' I would assume that many of the people associated with the Chomskyian perspective including Noam Chomsky himself have the same approach: that there's no point in observing something that certainly doesn't exist. (Dreifus 1998)

Let us now enter a more specific realm of the matter and go through some of the assumptions that seem not entirely accurate in "Questioning apes". On page 10, a severe position is taken against the ICE program carried out by Allen and Beatrix Gardner (whose essay appears on page 287 of the book). The methodological foundations of the project are called into question, as likely to produce

[48]As a general rule, not only in Gardners' case, no visitor is allowed inside the enclosures where chimpanzees are kept. Chimpanzees are some seven times stronger than humans, and they also have denser bones and thicker skin. In most instances of interactions, chimpanzees must restrain themselves to avoid hurting humans. When playing, chimpanzees normally throw, slap and playbite each other. However, if these actions elicit laughter among them, a human would be seriously hurt. The fact that they may master a bit of human language, unfortunately does not mean that they will use rethorics and diplomacy when the situation becomes critical.

"distortions", altering the results of the experiment, or—even more often—focusing on a wrong pattern or an irrelevant priority:

> One observer may prime another to inadvertently create situations in which a newly reported sign might be likely to recur by, for example, using a certain tool when the sign for that object has been reported by another observer. This outcome is especially likely owing to the fact that trainers were instructed by the Gardners to record the context in which a new sign was observed as well as the sign itself. Should the ape actually produce a sign in this sort of context, it would not be considered the result of outright prompting, molding, or the CHE, all of which the Gardners deny played a role in the reporting of new signs, and yet such innocent provision of opportunities for corroboration of other observers' records would certainly influence the overall course of the study. (Sebeok and Umiker-Sebeok 1980: 10)

Basically, the negative aspects would be two: encouraging—in whatever form—the production of a sign instead of another; and the presence, during the ICE, of contexts or situations that *facilitate* the emission of signs. The counterargument here is that this is exactly what human parents do: (a) they encourage the emission of certain words instead of others, even more insistently than the Gardners (as in the words "Mum" and "Dad"); and (b) they continuously create situations and contexts that facilitate the emission of signs and words (a recurrent instance being those language-related toys, such as picture books whose images are named in big fonts). It is to be inferred that this *is* a successful strategy for teaching language.

It also remains to be explained what is so wrong in *facilitating* the emission of signs. If—say—a dolphin D is trying to teach a human being HB the language of dolphins, HB would not really mind if D somehow encouraged him/her to emit 'dolphinesque' signs. In establishing how scientific an ICE program is, one should always take into account two main factors:

1. All the difficulties related to learning a communication system that is not only different: it is alien, it belongs to another species. It is much more difficult than learning a foreign, but still intraspecific, language; and
2. Simply enough, the trainee subject might just happen not to be a genius. Not only may Washoe be supported in the emission of human signs: she *must* be supported, as any person who is learning an alien language. After all, are we after teaching language to Washoe, or are we after teasing her?

The argument here is that the scientific cleanliness of any experiment should not be exercised purely for the sake of cleanliness. The target of the experiment is much more important than its aesthetics. If what we want to check is (a) whether non-human animals are able to communicate with human language; (b) whether they can ask what they need/want through the use of human language; and (c) whether they are able to model their perception with human language (as well); then our ICE program shall achieve a degree of methodological cleanliness that does *not* interfere with, or negatively affect, these targets. On the contrary, we shall create the most encouraging and inspiring environment for these targets to be reached. If a chimpanzee, or a specimen of another species, learns the human language despite all the obstacles that a semiotician claims are absolutely necessary for the scientific validity of the test, than what we prove is simply that the chimp is a genius, not that chimps are able to learn human language.

On page 14, again commenting upon the Washoe project, the Sebeoks agree with Eric Lenneberg (Sebeok and Umiker-Sebeok 1980: 80) in negatively evaluating Washoe's spontaneous ASL-signs production. If the Gardners look favorably at the creation of new signs in Washoe, as symptoms of enthusiasm towards, and cognitive appropriation of, the new communication tool, Lenneberg does not hesitate to apply Ockham's Razor and Morgan's Canon to such instances, and finds that "we are simply testing our own ingenuity to assign interpretations to productions that might, for all we know, have been emitted randomly".

The nature of these tools will be discussed at length later in this chapter: what however strikes one immediately is the fact that such a paradigmatically creative entity as language is here contested in its very nature. Language is by definition something flexible and imaginative: its codes—so useful for associating a given sign to a given meaning—are constantly challenged by neologisms, metaphors, loans from other codes, and other factors. When people speak a foreign language with native-speakers of that language and they have problems in finding the right word, it is precisely their creativity that rescues them from misunderstanding: they will use the word they have in mind in their own language hoping that it is somewhat similar to the other idiom, they will make gestures, they will point their finger towards something similar, they will describe that one word with an entire sentence, they will make up a new word if necessary… certainly they will come up with something, and in the end—rest assured—their interlocutors, although maybe a bit amused, will understand what they have in mind.

It is this very quality that makes language an always-in-progress entity. In Italy, there was no such thing like a straight transition from Latin to Italian. Uncountable hybrid idioms were created in the meantime, and even today what we call "Italian" is not entirely the same collection of words that it was only a few years ago.

Therefore, the spontaneous creation of new ASL signs is in fact to be welcomed as a very positive reaction from Washoe to the human language. She might not have learned the right sign for the right word, but there she goes trying to make her point anyway. And this is especially remarkable when the so-called "innovations" take place. I already mentioned some of these cases: an innovation takes place when the trainee does not know the sign for a given object, therefore simply makes up a new sign, based on a cognitive association (*Finger bracelet* for a ring, *White tiger* for a zebra, etc.). This is a rather convincing proof that language is used precisely as a modeling system, other than a mere communicative device. Of course, for someone who is at the same time opponent of the ICE and promoter of the conception of language as a modeling system, this is hardly acceptable:

> Unfortunately, these instances are generally reported in such a way that not enough is learned of the context of occurrence to enable us to rule out the possibility of either trainer suggestion or over interpretation. […] there is available a more parsimonious explanation for Fouts' report that Washoe was creating a new compound lexical item when she produced "water" plus "bird" in the presence of a swan and was asked, "what that?". Since Fouts provides no evidence that Washoe characterized the swan as a bird that inhabits

> water, we may just as well assume that Washoe, who was very familiar with the question form what that?, was merely responding by first identifying a body of water, then a bird. (Sebeok and Umiker-Sebeok 1980: 15)

Again, counterarguments are many, and, it must be stressed, *particularly* when it comes to this point:

1. It is simply untrue that the instances of linguistic inventions are badly or insufficiently reported. A few have already been reported here, and there are more. One has to simply check the relevant texts for confirmation;
2. Of all the possible examples, the Sebeoks chose to discuss one of the very few instances of linguistic inventions in which the interpretation is subject to ambiguity: "water" and "bird" are indeed rather general concepts, so it is in principle possible that Washoe could have referred to the body of water first, and to the bird afterwards. But then, what about "drink fruit" for a watermelon? "Drink" is an action, not a tangible object. What could Washoe possibly have meant, if not that kind of fruit that is actually so juicy that its consumption is more similar to drinking than to eating (at least in comparison with other fruit)? What about Koko's "white tiger", then? Did she see a "body of white"? And where did she see the tiger?;
3. Even considering the case of "water bird" alone, can we really accept the (Morgan's canon-inclined) explanation that Washoe, once asked "What that?", referred to two different objects, one of which is at least as big as a pool (if not as a lake, or even as the sea)? How did it happen that, all of a sudden, Washoe became so generic and scarcely-selective? Imagine: somebody takes Washoe out in the garden and points at a football, asking "What that?". Should we really expect that she comes up with something like "Grass ball", or "Garden ball"? And why are the birds that fly not called "Sky birds"?; in addition,
4. Was there nothing else than just a body of water and a bird? Not a tree, not the sky itself, not some clouds, not another animal, not a single element that might have caused Washoe to call the poor swan something like "water tree cloud bird"?; finally
5. The incidental comment "Washoe, who was very familiar with the question form *What that?*" should not go unnoticed. That is quite a statement, if one considers that Sebeok does not believe that Washoe (or any other language-trainee) does actually understand the questions that she is asked. His idea, as firm supporter of the CHE, is that Washoe reacts to that question by randomly making gestures, until she notices that her trainer is satisfied. How come she is now "very familiar" with the question? Could it just be that she understands it?

On pages 15–16, the Sebeoks' review focuses on the question of language manipulation for the purposes of amusement, what has been nicknamed "Monkey humor". In an often-reported dialogue, the ASL-trained gorilla Koko teases her trainer Dr. Francine Patterson, by pretending she wants her apple juice in the most unexpected places (nose, eye, ear, and finally mouth—accompanying all these jokes with a laugh). The manipulation of a code for personal amusement, as applied to non-human animals, is something that does not convince the Sebeoks, who again seem more at ease by calling into question the CHE:

> If a sign or other response produced by an ape appears to be inappropriate […], human trainers appear all too willing to stretch their imagination in order to make the animal's performance "fit" conversationally. […] Thus, anomalous chimpanzee or gorilla signs may be read as jokes, insults, metaphors, or the like, much as the not infrequent offenses against the very elements of counting and the fundamental arithmetical processes made by Clever Hans were regarded in part as intentional jokes and by an authority in pedagogy as a sign of independence and stubbornness which might also be called humor. (Sebeok and Umiker-Sebeok 1980: 15–16)

Further on, that very example of Koko's humor is reported as a typical CHE-affected misinterpretation: the truth, it is maintained, is that Koko was simply unable to perform the correct sign for "mouth". As before, a number of remarks can be made:

1. As the request for food or drink is by far the most recurrent ASL interaction between Koko and Patterson, it is extremely unlikely that Koko has not mastered it properly, and that this episode is something other than just an amusing diversion to a routine action. In addition, Patterson in any case reports other cases of her gorilla's humor, like when she was asked "to place a toy animal under a bag, and she responded by taking the toy and stretching to hold it up to the ceiling" (Patterson 1978: 456);
2. These examples show a typical and basic feature of humor: turning things upside down. A asks something to B, and B does the exact opposite. B has fun by teasing A, and teases A by having fun. Nothing is more elementary than this, in humor;
3. If Koko was just making mistakes, why on earth would she laugh?

Staying with Koko, and signs of "independence and stubbornness", there is another critical remark on page 37. In commenting on one of the ASL tasks proposed to Koko, Patterson (1977: 10) noted that the gorilla was not particularly attracted by the exercise, and after a while (five trials per day, maximum, and two sessions per week) she would stop being co-operative and start performing the same sign over and over, or not performing any sign at all, or generally displaying boredom. Excluded these instances—said Patterson—Koko was able to perform the right signs in 60 % of the cases. The Sebeoks claim that this method of counting was incorrect:

> Koko is said to have scored correctly 60 % of the time, or above chance, on the series of double-blind tests which were administered to her. It is reasonable to ask, however, how Koko's instances of avoiding the test, e.g., by responding to all objects with the same sign, were scored. We are not told whether or not such inappropriate responses were discounted as avoidance measures or counted as errors. […] We can only guess whether or not Koko's performance would have been below chance were a less biased accounting to have been made. (Sebeok and Umiker-Sebeok 1980: 37)

From a general understanding of Patterson's research, the Sebeoks are certainly right in suspecting that she did not count Koko's reluctance as a mistake. But the question is, why should she have? Why should the lack of enthusiasm to perform a given task be equaled to a wrong performance of it? Let us compare the situation to an ordinary human instance of two people (X and Y) playing a game, for example chess. As the game proceeds, X makes both good and bad moves: in one instance X successfully captures one of Y's castles, on the next move X makes a mistake and loses a knight, and so on. At some point, 60 % of X's moves have proved to be correct, and 40 % have favored Y. So, overall X is leading in the game. Except that, well, X gets a bit bored and wants to quit. Y, however, insists that they continue, despite X's attempts to persuade Y to the contrary. X therefore refuses to make any further moves, but again Y insists. The two go on like this for a while, until X gets a bit irritated and decides simply to move the queen back and forth until the game is lost and an end can finally be put to it. At last, Y understands and allows X to leave.

Now, X's attitude can certainly be criticized from several points of view—most of all for lack of politeness—but no one could really say that X's reluctance to play corresponded to a series of *wrong* moves on the chessboard. Not even when the queen was moved redundantly back and forth: it was very evident that X was not interested anymore, and was just trying to persuade Y to finish the game. An ideal observer recording the successfulness of the two people's game will definitely agree that X's percentage of good moves is still 60 %, and that the final sabotage does not really count. I certainly know this as a parent of a small child. If my son and me engage into some educational game (like forming words with letter cubes), it is extremely easy for me to tell the difference between a word wrongly composed, and one that he on purpose assembles randomly because he wants out of the game and back to his beloved toy cars. I do not count the latter instances as mistakes: I only count those right and wrong words that were formed *before* he got bored.

On page 16, the Sebeoks' target is mistakes and imprecision:

> The determination of trainers [the Premack's team, this time] to show their animal [the chimpanzee Sarah] in the best light has even led to the denial that errors are mistakes, again bringing to mind one of the assumptions which sets research on psychic phenomena apart from normal scientific procedures. The so-called psychic's mistakes on tests of her powers are frequently used to prove that those powers are real, based on the assumption that if the performer was using mere tricks she would be correct every time. (Sebeok and Umiker-Sebeok 1980: 16)

Once again, there can be a line of reasoning that make Premack's attitude perfectly logical: if the trainee of a given experiment performed all his/her tasks without a single error, one really would think that s/he had found a way to cheat. If however the performance displayed a minimal percentage of mistakes, one would be more reassured that the ape, who—it must be assumed—is more likely to be normally-gifted, intellectually speaking, than a genius, has done his/her job in the best possible way.

Besides, the point is that the percentage of tasks performed correctly was significant, (i.e., more than 60–70 %): if one answers correctly 6–7 questions out of 10 in any school exam, one will normally pass the exam.

Staying for a moment with recollections of school, it is difficult to agree with a remark made a few pages further on:

> A related issue concerns the personal preferences exhibited by the apes for certain project members. All the animals used so far in the ape "language" projects have shown some preference for certain trainers, performing better for these "favorites" than for the rest, but there has been no systematic attempt to account for these differential responses on the part of the animals. (Sebeok and Umiker-Sebeok 1980: 22–23)

What seems to be forgotten here is that the trainees are first and foremost subjects, with their characters and preferences. They cannot be the infallible machines that the Sebeoks are demanding in order to take their learning-efforts seriously. We all have our personal recollections of school subjects that we liked, but we were not able to learn properly *precisely because* certain teachers were not kind to us. The converse is also (and more significantly) true: there were subjects we did not like, and yet we managed to learn well thanks to teachers who were both kind and competent. Not to mention that more often than not, our professional choices (especially academic ones) become clear only when we find a true mentor, who understands our potential and points us in the right direction. In other words, having preferences for one trainer instead of another is not only normal and scientifically-acceptable. It is also healthy.

Incidentally, it is also mentioned (Sebeok and Umiker-Sebeok 1980: 23) that the trainers, too, have their own preferences, among the trainees. Again, do we not remember our schoolmates who were *teacher's pets*?

One final remark in the "Questioning apes" text calls for some further reflection —the observation on page 30 that every now and then, the apes use signs from their own non-human repertoire in order to support their ASL, which results in the observer being "unable to determine which of the actions he sees performed by such an animal are part of its natural repertoire and which are the results of special training" (Sebeok and Umiker-Sebeok 1980: 30). Once more, it is difficult to understand why this should be a problem, when in fact it could be regarded as a bonus. It shows active (possibly enthusiastic) participation of the trainees in the communication process, and—most of all—it shows very clearly that they have fully understood that it *is* a communication process. That is why, when unable to make themselves clear by means of the human code, they do not hesitate to use their own sign system, hoping, perhaps in vain, that their interlocutor will finally get their point.

In concluding this section, it should not surprise anybody that Sebeok, the man who so convincingly argued in favor of the fully-semiotic qualities of non-human communication is here so much against the extension of the notion of language to other animals. We cannot be surprised because his definition of language has always been very clear, as something whose communicative characteristics are not at all important, in comparison with its features as a modeling system. Therefore, talking about communication is an entirely different issue than talking about language.

What perhaps *is* surprising is a certain inaccuracy and (what seems to be) an active prejudice in his arguments against the ICE, as if he has already made up his mind and is not going to change it, whatever happens. To sum up Sebeok's arguments in a few words, it turns out that, for him, the true learning of a language entails the following conditions:

1. the avoidance of any encouragement or support during the learning period;
2. the prohibition of creativity and imagination in producing signs;
3. code manipulations must be understood as linguistic mistakes;
4. boredom and insubordination are taken as linguistic mistakes;
5. the acceptance only of instances of communication where not a single error is made;
6. the prohibition of any form of personal preference towards any trainer;
7. the prohibition of the use of one's own original sign repertoire to support the learning of the new idiom.

In such a learning environment, one must conclude that it would not only be Washoe who would fail to progress in human language studies, but even Umberto Eco. Not to mention that growing a small child in these learning conditions would produce enough arguments for the authorities to take the kid away from such a cruel family.

One last general note about the CHE. Even if, for the sake of argument, one accepted that the ICE are CHE-biased, the fact is, there is a very thin line between imitating and understanding. Even imagining a fully-CHE-biased interaction, it is hard to find a significant difference between performing a sign because "this is how I get the apple", and performing it because "I know this stands for 'give me an apple'". As a matter of fact, both approaches prove that the trainee has learned how to achieve what s/he wants through the code s/he is expected to use. As soon as s/he has more than one sign (say, just to be banal, A for apple, B for banana, and C for candy), and processes the information in such a way that s/he performs B when s/he wants a banana, and not when s/he wants a candy, then the task is successful, whatever the exact configuration of his/her reasoning is, either it goes like "I will perform B, so I get a banana", or like "I will perform B, which stands for banana". The only difference is that the former relation is indexical, the latter is symbolic, but the connection sign-object is there anyway, which is exactly what the ICE detractors deny. We might certainly accept that a conative type of communication proceeds by indexes, especially when the trainee proves s/he can use this foreign code in a symbolic way as well (I would in any case hope that by now, enough evidence in that respect has already been provided, here).

3.3.3.5 Language and Umwelt

Another aspect of the Sebeoks' review that somehow surprises is that they seem to care a lot about aspects that are probably not that important. The aim of this paragraph is thus to point out which ICE-related issues have been either underrated and/or insufficiently dealt with. In that respect I shall not spare the ICE scholars themselves a certain amount of criticism. The claim here is that several aspects involved in these projects were overlooked or misunderstood. Now, that does not apply only to Sebeok and other semioticians, but to the scientists involved in these projects too.

To start with, even though many of these scholars seem to feel genuine love and compassion towards the non-human animals they are dealing with (which is certainly the case with Roger Fouts, Sue Savage-Rumbaugh and Francine Patterson, to mention some of the cases I am most familiar with[49]), it must still be emphasized how ethically questionable it is to propose that a non-human animal may be stolen from his/her own environment and put in a cage for most of the time, forced to establish relations and perform tasks that s/he probably does not want or need, and exposed to a great degree of boredom and depression, if not—unfortunately—violence and deprivation.[50] It is no chance that the most successful ICE (Washoe, Koko, Kanzi, Alex, etc.) were those where the trainees were given the best possible quality of life.

Apart from this, there are other issues that probably deserve more careful scrutiny. One of those certainly concerns the application of the Umwelt theory (which I have briefly introduced in a previous footnote). It has been said several times (including by Sebeok himself of course) that belonging to a different Umwelt is a sufficient reason for a non-human animal not to be able to access a sign system that relates exclusively to the human Umwelt. And on the other hand, it has been argued that the human Umwelt, *unlike other animals* (here we go again!), is able to access any other Umwelt, at least in principle, thanks to the inherent characteristics of language itself, in particular what I have previously called "Narrativeness". Also, if one considers, somehow, past and future (especially future) as 'possible worlds' (after all, possible-to-happen future or possibly-happened past are *things* that are not yet or not anymore *objects*), then also the quality that was here called "distant space-time semiosis" belongs to this discussion.

Moreover, even if this has been emphasized far less often, there is definitely an intrinsic difficulty in configuring an Umwelt that is different from one's own (and that applies both sensu stricto, i.e., accordingly to Uexküll's formulation of the concept, and sensu lato, for instance when we talk about cultural or personal Umwelten). Now, even though most of these statements are difficult to deny, they

[49]And those who also happen to have the greater media exposure.

[50]An entire chapter of Roger Fouts' beautiful best-seller *Next of Kin* (1997) is devoted to the cruel conditions in which Washoe had to live during the period she moved from the Gardners' house in Nevada, to William Lemmon's laboratory in Oklahoma.

can still be reformulated in a more accurate way. Firstly, as already mentioned, it is certainly true that the decoding of alien Umwelten is almost a contradiction in principle. Taking Uexküll's theories very literally and logically, one might deduce that if we, members of the Umwelt A, manage to decipher the Umwelt B, then we cannot really talk about different Umwelten anymore, because the occurrence of this very process would prove that the Umwelt B is simply part of the Umwelt A, therefore not 'alien' (and in fact not Umwelt). When we talk about *things*, in a way, they are no longer *things*, but already *objects*, or at least in a philosophical sense.

Secondly, the Narrative and Distant Space-Time semiosic qualities of language (the former, in particular) seem to create a certain uniqueness in the human Umwelt, in that they allow a strong dialectical, philosophical and even rhetorical configuration of alien Umwelten. *But*: dialectics, philosophy and rhetoric are not enough to ensure *full* access to another Umwelt, even if we can make some significant observations. As already said, particularly from an ethical point of view, this should not be forgotten, as we like to think that, when we investigate other animals, we are able to understand *all* that is there to understand. How does it feel to live in a world dominated by olfactory stimuli, rather than visual ones? Are we sure language is enough to fully grasp that? And what about channels that are completely alien to our semiosis, like the electric one? *How does it feel to be a torpedo?*

More importantly, now that we know that the ICE, in most cases, turned out to be successful attempts to teach human language to other animals, we need to ask ourselves a few questions concerning the supposedly *qualitative* differences between our human Umwelt and others. If language supposedly belongs to the human Umwelt, then it follows automatically that Kanzi & co. *did access* the human Umwelt, from the front door (that is, through the feature that by definition characterizes the human species). Mastering language, in the extensive way humans do, leads to increasing narrativeness and distant space-time semiosis, therefore the more refined and articulated our linguistic knowledge, the deeper our narrative abilities. However, the outcomes of the ICE call all this into question, probably preventing us from claiming a qualitative *exclusivity*. If anything, they prove that narrativeness and distant space-time semiosis are language-specific features. But that is an entirely different matter, because now the difference is between language and other communication and modeling systems, not between humans and other animals. If other animals were able to learn language, at least to some extent, that means that access to more sophisticated semiosis is clearly within their cognitive potential.

This is not, by all means, an attempt to dismiss the Umwelt theory (please, make no mistake on this point). This is a warning on what kind of features we decide to put into the Umwelt-box, and if we have a valid scientific reason for such choices. In fact, the point is that it is not so much "features" that characterize Umwelten, but rather experience, subjectivity and interpretation processes.

3.3.3.6 Ability of Language Versus Need for Language

In the light of all this, the big question is: why, as a matter of fact, did language develop in humans and not in other species (at least we do not have significant evidence to prove otherwise)? And why does language not arise spontaneously in other animals (since they were able to acquire it only after human's guidance, in contexts of captivity)? Obviously, those questions deserve a separate and thorough treatment, and a whole different set of theoretical and empirical work. Nevertheless, please, let me mention one or two reflections, here, in order to make my argument a bit more complete, and in order—once more—to attempt a numanistic (semiotically-informed, in my case) way to discussions of this sort.

My claim (which counts for nothing more than "my claim") is that language arose in the species that more than any other *needed* it, not in the only one that *could* invent it. The existence of some 3000 different idioms in human communities led to the common place that language is a purely cultural phenomenon. In actual fact, no less than forty universal phonemes exist, and a selective attention towards them among infants from all over the world has been empirically observed, proving the spontaneous, thus biological, basis of symbolic language (see Malacarne 1992: 452). To accept this point makes it easier to accept that language must have been born as a result of an evolutionary need.

One opinion (see also Malacarne 2005: 243–247) is that the need for a language came as a solution to the many physical limitations of the human being and as a consequence to a few advantages. In most of the problems that nearly any other animal can solve alone, or at least with simple actions or decisions (feeding being the most obvious example), human beings need cooperation. In that respect, language could have been subsequent to strictly adaptive functions of communication in general, such as the referential or the conative one, which required more complex specifications, due to the inability of humans to perform a given task on their own.

In particular, the referential function (this one, too, along with the conative, being found in many animal species) had a fundamental adaptive value (see also Cimatti 1998: 71–72), in that it allowed a true body extension to times and places that were not directly attended by the interlocutors (a human or other animal can be *informed* of the presence of food in a certain area where s/he has not been, or is not at the very moment). The more efficient and articulated the communication, the more successful the cooperation (as one can also see in species like whales or wolves, both deciding the hunting strategy before actually putting it into action). A signal arose with a specific function, e.g., to obtain from the receiver a given tool that the latter had to go and collect from another place. Later in time, such a function must have developed in more articulated ways (passing through stages of generalization, abstraction, etc.), going hand in hand with natural selection and phylogenetic evolution of the species. The most successful human subjects were those who could communicate in the most precise and articulated way. Operatively speaking, the very *selection* consisted in the development of two brain areas, the Broca (that controls the motor expression of sounds) and Wernike (that is involved in the perception of sounds), and—above all—the vocal apparatus. When scientists say that we are similar to Chimpanzees for around 98–99 % of our genetic heritage, they normally fail to add that within that 1–2 % of difference there is precisely the vocal apparatus and these two brain areas. The causal relation between these elements and the creation of language is reciprocal: the former were responsible for the development of the latter, and the latter was responsible for the development of the former.

And it was of extreme importance to have that kind of resources in order to develop a device such as language: the vocal apparatus, in particular, made it possible to exploit the richness, the flexibility and the practicality of the vocal-acoustic channel. In other words, having such a flexible organ for speaking, human beings, literally, *spoke*. A cognitive use of the language was thus made possible, and the process of extension and enrichment began, evolving into the virtually limitless articulation of nowadays: it was during this process of expansion that—one may guess—language became more important as a secondary modeling system. The very same process, as applied to hands, is described by Friedrich Engels to tool-fabrication and work (1896: 545–554): hands became at the same time the cause and the product of work.

3.3.3.7 Another Plea for Fairness

Related to this issue is also another, quite crucial, consideration: so far, the ICE have been evaluated only at an ontogenetic level, and not at a phylogenetic one. When we think about the transition from the birth of the genus *Homo* (4–5 millions of years ago), when the first—very rudimental—traces of a language presumably appeared, to the *Cro-magnon* (125,000 years ago), with whom we probably have the appearance of a less-embryonic form of language, we understand that it took millions (!) of years for language to take a recognizable shape. The above-mentioned anatomical and neurological features developed at an extremely slow pace (and randomly, if one takes a strict Darwinian perspective): on the contrary, non-human apes were introduced to language, in a methodologically proper way, only half a century ago, with very few subjects, and in captivity. How can we reasonably expect perfection already from this first generation of language speakers, and describe the circumstantial difficulties as "failure of the whole enterprise", when we ourselves needed millions of years to shape language? Plus, even if at some point (in who knows how many centuries) we witness in apes linguistic capabilities comparable to an average human adult, we would still have to see whether, reintroduced in the wild, these subjects would benefit from such competence, that is, if their species *needs* language. Finally, if we find out that they do, we will have to wait another shot of who knows how many millennia before natural selection creates a language-inclined non-human species. Now, with all due respect, *this* is a fair, non anthropocentrically-biased, line of reasoning.

Anyway, origins of language aside, what still remains is my claim that narrativeness and distant-time semiosis are not features that characterize exclusively the human language. In that respect, a "qualitative" difference between human language and other communication systems would probably appear to be the existence of these "linking signs" mentioned earlier, that is, those signs whose exclusive function is that of connecting signs that produce an actual meaning. Curiously enough, the big advantage of linking signs, and language as a consequence, is not the ability to make meaningful signs, but exactly the opposite, i.e., creating meaningless signs.[51] Signs that do not refer to anything when taken alone (unless a specific metaphoric use is made of them, but that is an entirely different issue), but can make a great deal of difference when combined together with meaningful signs, in certain numbers and sequences. Charles Hockett, in defining the famous (some would say infamous) sixteen design features for human language (1960), had a very similar concept in mind when describing the 12th feature, duality. Non-human semiosis seems to operate most of all in economic and ergonomic modes: signs, when used to communicate, must all make sense. Roughly speaking, a non-human animal emits five signs if s/he has five things to say. Humans may employ a great number of signs just for expressing one thing. But it is exactly this 'meaninglessness' factor that increases exponentially the semiotic potential of human language: it creates specifications, differences, separations, distinctions, and so forth. In a word, it creates a virtually unlimited number of relations.

So, must we finally surrender to the anti-Darwinian notion that, after all, there is some *qualitative* difference between humans and other animals, even if by now this difference is reduced to articles and propositions only? Of course not. In this case, too, it would be scientifically too risky to launch a new "ultimate sign of human distinction" campaign, and my point is exactly that we have to learn to be careful with that. Enough of that mistake.

Plus, the fact is, there *are* cases, within the non-human animal world, in which an apparent anti-economic use of communication occurs, and where the sounds employed do not necessarily correspond one by one to a specific meaning, while the entire formulation of a phrase *does* say something. It is, predictably, the case with birds and cetaceans (animals with very rich communication repertoires), particularly those instances that we call songs. A species like the humpback whale, *Megaptera novaeangliae*, is a very good (and notorious) example. Humpback whales are passionate manipulators of their compositions. They provide their songs with continuous variations, so that, after about five years, they sing a totally new song and, as far as scholars could notice, they never come back to the original version. These variations occur during the mating season. A phrase may slightly be shortened or slightly prolonged, or suddenly interrupted and replaced with a new

[51]Some may raise the objection that there are languages which in fact do *not* have linking signs, and replace them with different declinations of nouns. But this is not really relevant, since the "linking" parts will simply be inserted in other words. If I want to say in Lithuanian (one of such languages) "University of Kaunas", I will not have an equivalent of the linking sign "of" as a *separate* word, but I will need to intervene on the word "Kaunas". I will therefore say "Kauno", as genitive of Kaunas, and Lithuanian speakers will understand the equivalent of "of Kaunas". And even in these languages, some linking words still exist, in fact. Lithuanians still have such words like "su" (with), "be" (without) and others.

one. A theme may be enriched with new sounds and then inserted into the whole context, or a part of it can be changed in sequence. Every whale updates and includes new melodies in its song. They are performed faster and the variation is kept within the song until the next mating season. Then, old songs are reprised at the very point they had been left last year.[52]

Now, there are certainly no elements to affirm that those extra meaningless sounds emitted by a whale are something qualitatively comparable to some human "of", "the", "with", "to", i.e., to "linking signs". The evidence is however that there are cases (and not only a few) of non-human animals using communication in an anti-economic way, cases, that is, where they say more than what they mean. Not to mention (although it does not count for the type of argument that I am developing here) that anyway language-trained animals easily managed to learn linking signs like "this/that" (Fouts 1998: 293) or "on", "and", "in" and "of" (Mainardi 1975: 125–129), For what we know, anyway, human language remains a highly specialized form of communication in this very respect.

Coming back to the issue of Umwelt, in the strict sense, there is another issue that both ICE researchers and semioticians seem to have overlooked, and that is the importance of the way the perceptual field of a given specimen (therefore, probably, of an entire species) affects the actual reception of a linguistic sign. An interesting observation that was made on the Washoe project was that, once learned the ASL sign for *flower*, she took to name that way everything with a strong smell, therefore using a form of iconic categorization that is not visual, as it would be almost systematic in a human being, but olfactive:

> When FLOWER started to become a common item in Washoe's vocabulary, it was reported in several inappropriate contexts that all seemed to include prominent odors: for example, Washoe signed FLOWER when opening a tobacco pouch and also when entering the Gardner kitchen at a time when there was a chicken boiling on the stove. Taking our cue from these observations, we modeled and molded the SMELL sign in appropriate contexts. For some time, FLOWER [...] persisted as an error, but gradually, a highly reliable, well-formed SMELL sign emerged. (Gardner et al. 1989: 82)

Washoe's signing must have certainly looked "inappropriate" to the specimens of a species which bases so much of its perception on the visual channel. We are given a flower, and we are told "this is a 'flower'": in 99 % of the cases we will recognize other members of the category "flower" on the basis of a visual resemblance: a picture of a flower, a painting of it, a plastic flower, and maybe even an object that is not designed to resemble a flower, but—as *it kind of looks like one*—we can still label it as such in order to clarify what we mean to an interlocutor. Washoe obviously reflected in a totally different manner (and that, once more, should act as a reminder of the importance to fairly consider Umwelt differences across species). The relevant characteristic of that object, to her, was its smell, i.e., the olfactory feature of the flower. Consequently, the semantic field called "flower"

[52]For an extensive description of the concept of variation in zoomusicology, see Martinelli (2009: 145–153).

became to her the field of smelling objects, no matter what their visual appearance (as for humans the semantic field of flowers is a field with certain visual features, no matter what their olfactory appearance). A bottle of perfume, in this sense, fits more to Washoe's semantic field of "flowers" than, say, a Van Gogh's painting of sunflowers. Nothing, in this process, is less "clever" or logically-legitimate than the type of categorization established by humans.

This instance is not isolated. A similar episode happened when Washoe, who was also one of the chimpanzee-painters (see Desmond Morris's observations in Morris 1962), was asked to draw a ball. Again, as humans, our first idea for representing a ball is iconic: we trace a circle on the paper, and perhaps—if we are football fans—we add black pentagons and white hexagons inside it. But to Washoe, it seems, the most relevant iconic characteristic of the entity 'ball' is kinetic, not visual, as she drew a series of progressively smaller arcs, i.e., the trajectory of the ball. Once more, such a reasoning process is perfectly acceptable.

In this respect, the fact that many of the language-trained non-human subjects have proved in several different cases a high degree of flexibility in accessing the human Umwelt, and reconfiguring their own accordingly (for instance, accepting the symbolic representation of a series of objects that they would usually represent indexically), shall be saluted as one of the most remarkable achievements of these ICE.

3.3.3.8 Denying the Subject

Finally, however banal it may seem, it is important to emphasize that the identity of the ICE trainee as an individual, and not as a member of a given animal species, is by all means crucial, but has been heavily underrated by both sides of the discussion on the ICE (researchers who performed these experiments, and scholars who commented on them). The fact is: in all these experiments, human language was not merely taught to "a species", but first of all to individuals named Sarah, Washoe, Moja, Lana, etc. Koko and Mike are first of all Koko and Mike, and only then they are gorillas. And Kanzi is first of all Kanzi, and only then he is a bonobo. This list of tautologies is quite important here, since it reminds us of the following things:

1. The trainee is first of all someone who has an individual (cognitive, emotional, etc.) profile, and is exposed to experiences and events, *then* s/he is someone on whom an ICE program is performed;
2. The individual responses of the trainees may be related to factors that are external to the experiment's environment, and may simply be the result of a

personal experience, or be part of the cognitive-emotional background of the subject in question; in particular

3. Observations and results (either negative or positive) of the ICE have often been inappropriately generalized to an entire species, at least, if not to entire orders, or to the whole Animal Kingdom *tout court*. In simple words, when Moja fails to perform the right ASL gesture, it does not mean that all chimpanzees have failed. Likewise, if Moja will someday compose a rhyming sonnet in ASL, that does not mean that we should expect the same from Lana, Lucy, Sarah and the others;
4. An extensive account of the personal history of the subject should always be integrated with the rest of the ICE, as a fundamental and unavoidable element for the correct interpretation of the trainee's response. The success of Fouts and the Gardner's with Washoe and other chimps, and of Patterson with Koko and Mike, is possibly due to the fact that this aspect was fully taken into account. As an example, one may read how carefully Fouts (1997) relates Washoe's progresses with ASL to specific events (positive and negative) in her life.

Still, many seem to forget these apparently basic notions. Perhaps, the most evident instance where the individuality of the trainee is totally ignored concerns our expectations towards his/her behavior. It is what we may half-jokingly call "thirdness-only expectations", with obvious reference to Peirce's notions of Firstness, Secondness and Thirdness. We expose a non-human animal to a problem-solving task, by—for instance—introducing a variable in his/her usual environment. We do not only expect the trainee, who *does not* know s/he is supposed to perform a task, to focus quickly and only on that very variable, but, while doing that, we also expect him/her to privilege a line of reasoning which is by no means the most obvious and spontaneous one.

This so-far-ambiguous concept can be exemplified through the words of someone who was able to depict it with extraordinary efficacy. Sometimes, in fact quite often, philosophers and artists are able to see much further than scientists, and this is probably because, most of the times, they do not see empathy and sensibility as biases (talk about humanities as "added value"). They rather use them as devices for a better comprehension. In my humble opinion, no one has succeeded better in picturing the paradoxical situation just described than the writer and Nobel-prize winner John M. Coetzee. Apologies for the long quotation here, but it really makes the point clear. Plus, it is a great piece of literature:

> Let me recount to you some of what the apes on Tenerife learned from their master Wolfgang Köhler, in particular Sultan, the best of his pupils, in a certain sense the prototype of Red Peter. Sultan is alone in his pen. He is hungry: the food that used to arrive regularly has unaccountably ceased coming.
>
> The man who used to feed him and has now stopped feeding him stretches a wire over the pen three metres above ground level, and hangs a bunch of bananas from it. Into the pen he drags three wooden crates. Then he disappears, closing the gate behind him, though he is still somewhere in the vicinity, since one can smell him.

> Sultan knows: Now one is supposed to think. That is what the bananas up there are about. The bananas are there to make one think, to spur one to the limits of one's thinking. But what must one think? One thinks: Why is he starving me? One thinks: What have I done? Why has he stopped liking me? One thinks: Why does he not want these crates anymore? But none of these is the right thought. Even a more complicated thought—for instance: What is wrong with him, what misconception does he have of me, that leads him to believe it is easier to reach a banana hanging from a wire than to pick up a banana from the floor?—is wrong. The right thought to think is: How does one use the crates to reach the bananas?
>
> Sultan drags the crates under the bananas, piles them one on top of the other, climbs the tower he has built, and pulls down the bananas. He thinks: Now will he stop punishing me? The answer is: No. The next day the man hangs a fresh bunch of bananas from the wire, but also fills the crates with stones so that they are too heavy to be dragged. One is not supposed to think: Why has he filled the crates with stones? One is supposed to think: How does one use the crates to get the bananas despite the fact that they are filled with stones?
>
> One is beginning to see how the man's mind works.
>
> Sultan empties the stones from the crates, builds a tower with the crates, climbs the tower, pulls down the bananas. As long as Sultan continues to think wrong thoughts, he is starved. He is starved until the pangs of hunger are so intense, so overriding, that he is forced to think the right thought, namely, how to go about getting the bananas. Thus are the mental capabilities of the chimpanzee tested to their uttermost.
>
> The man drops a bunch of bananas a metre outside the wire pen. Into the pen he tosses a stick. The wrong thought is: Why has he stopped hanging the bananas on the wire? The wrong thought (the right wrong thought, however) is: How does one use the three crates to reach the bananas? The right thought is: How does one use the stick to reach the bananas?
>
> At every turn Sultan is driven to think the less interesting thought. From the purity of speculation (Why do men behave like this?) he is relentlessly propelled toward lower, practical, instrumental reason (How does one use this to get that?) and thus toward acceptance of himself as primarily an organism with an appetite that needs to be satisfied. Although his entire history, from the time his mother was shot and he was captured, through his voyage in a cage to imprisonment on this island prison camp and the sadistic games that are played around food here, leads him to ask questions about the justice of the universe and the place of this penal colony in it, a carefully plotted psychological regimen conducts him away from ethics and metaphysics toward the humbler reaches of practical reason. And somehow, as he inches through this labyrinth of constraint, manipulation, and duplicity, he must realize that on no account dare he give up, for on his shoulders rests the responsibility of representing apedom. The fate of his brothers and sisters may be determined by how well he performs. (Coetzee 1999: 126–127)

Leaving aside the specific case of Köhler and Sultan, what matters here is the line of reasoning. Hardly, if ever, the emotional condition of the trainees, as subjects, is considered important. We do not expect Firstness from them: we have a straight expectation for Thirdness. We expect them to react to our stimulus in the way we have planned (and received funding for), while the rest is either uninteresting or wrong. Why? The same principle applies to many of the previous critical remarks on Sebeok: why should trainees not have their own preferences? Why should they not be supported in the learning process? Why should we expect them to be language-machines when they also have their lives and their personalities, other than a predisposition or none to learn ASL or lexigrams, or whatever they are?

3.3.4 Conclusions

From the amount of problems I have placed on the imaginary table of these two case studies, it may be a bit challenging to find a way to wrap them up into a consistent and not-too-dispersive set of conclusions. At the same time, it is clear that—from both a scientific and ethical point of view—there seem to be a great "narrative" behind this whole academic enterprise of establishing (in various forms) the radical, discontinuous, difference between human beings and other animals. There is a red line across notions like the semiotic animal, the threshold of language, the fear for anthropomorphism or for the CHE, the various manifestations of anthropocentrism and speciesism. It is a red line that, via the basic application of Morgan's Canon to animal studies (do not use complexity to explain animal behavior, when you can use simplicity), invokes the privilege of cognitive complexity only to humankind. All that follows (in all its manifestations) reflects this basic scheme. The semiotic animal claims a more complex semiotic behavior for the human being, and leaves the simpler one for the other animals; the critiques to Clever Hans emphasize that Hans could not be complex, but had to be simple; to interpret animal behavior in an anthropomorphic way means to see it as more complex, etc., up to building that basic dualistic cosmology that we never really got rid of: nature (simple) versus culture (complex); body (simple) versus mind (complex); animal (simple) versus human (complex). For the purposes of these concluding reflections, I suggest we use the word "morganism" to summarize the totality of the various reflections inspired by Morgan's Canon, starting of course from the canon itself.

I therefore suggest starting to define this red line by emphasizing, with the tools of humanistic reflection, the main contradictions of morganism and morganists:

1. When morganists talk about the 'possibility' of interpreting animal behavior in a given way, they are referring to those cases where a *speculative* choice is necessary, that is, where a non-human behavioral pattern that is analogous to a human one, does not give empirical and unquestionable evidence that it is also a homologous pattern. In other words, they refer to those cases where the animal *seems* to behave or think in a certain way (similar to the way a human would behave or think in the same situation), so the point is: does the animal *seem* to behave, or is s/he *really* behaving like that?

In these instances, evidently, all we have is this similarity between the human and the non-human behavior. Thus, logically speaking and when counterproof is missing, the starting hypothesis should be: the two patterns *are* homologous, precisely because they are similar, and that is the only thing that we know at this point. Then, we can pursue the necessary kinds of research, and try to understand if the hypothesis is confirmed or not. Still, at the very beginning, and "the very beginning" is the research phase to which morganism usually refers, we shall think of a homology (by "very beginning" I also mean that whole bulk of humanistic speculations upon these issues, which—within the context of the scientific method—correspond to hypotheses advanced *before* empirical evidence).

Let us take an example: on one side we have a car, on the opposite side we have an object that has a slightly different shape, that *seems* to move, *seems* to have four wheels, *seems* to have seats, a steering wheel, windscreen wipers and a car-stereo. At the moment, that is all we know. What are we going to do? Perhaps, we should start from the idea that this mysterious object is actually a car, maybe from a different brand than the one we have on the other side, but still a car, and our research should be oriented in that direction (are those really wheels? Are those really seats? Is there an engine, a brake pedal, etc.?). That seems to be a more convenient procedure than to think that the object we are observing is actually a skateboard, so those things that look like seats are actually a simple flat board, and the wheels are much smaller than they seem.

Still, this is what morganism does. Not only does it suggest that we should not trust appearances, it also tells us that reality is simpler (or more reduced) than it appears to be. Felice Cimatti (1998: 147–151) proposes an interesting and somewhat ironic example to illustrate the applications of this canon to the case of Washoe. According to the morganists, at the very moment Washoe is asked to perform the ASL gesture corresponding to a given object, she does not have the slightest idea of how to perform such a task, since she is not able to understand the relation between signifier (the ASL gesture) and signified (the object). Thus, she proceeds by attempts, moving her arms at random, until—unconsciously or not—her trainers display apparent satisfaction, which lets her know what the right gesture is. Here is a typical example of the CHE, which morganists find the most suitable interpretation. Unless one is able to invalidate any risk of the CHE, they say, the ICE are a waste of time. Most of all, the more or less emotional relation established between researcher and animal should be avoided by creating an "emotion-free" context. Of course, to consider the use of a communication system as unrelated to social and emotional interaction is a self-evident contradiction and a more serious mistake than anthropomorphism. What is the point in learning a communication system if you have no one to use it with, do not receive adequate stimuli, and, in the end, do not have any reason to learn it? In such a case, says Cimatti, the mistake is not only methodological, but also theoretical: "if language is a system based on emotional interaction […], then it is simply impossible to study it without considering this aspect" (translated from Cimatti 1998: 149).

In other words, here we face the type of situation that Paola Cavalieri (1999: 24–25) warns us against: the aprioristic parsimony of morganism requires very often more complicated explanations for the phenomena it comments upon, therefore contradicting its seemingly economic philosophy. In this way, inevitably, something of that object we were observing in the example remains unexplained. For instance, what is that thing that looks like a car-stereo? What about those look-alike windscreen wipers? How do they fit in with the idea of a 'skateboard'? Scholars in animal studies invented an ad hoc solution to these disturbing problems: a big, fat black box where they put everything that cannot be explained. The most common name for this black box is "instinct", possibly the most effective semantic field scholars employ to keep non-human cognition on the "simple" side of the above-described dualism. The idea, indeed, is that we can describe reason and rationality as "complex" categories, and instinct as the simple one.

Although seemingly the most easily definable concept, at least in non human-related contexts, instinct is in fact the most complex and tricky one. Decades of lively discussions among scholars have not yet produced even an approximate definition. Instinct has been defined as a "voluntary but never-previously-put-into-action behavior", a "behavioral impulse for the accomplishment of biological functions", a "stereotyped behavioral pattern", and a "pattern put into action without a precise idea of its results". It is already evident that these definitions are more hiding than showing: they just describe an action, but they do not give account of causes and articulation. Moreover, some of them are in contradiction with each other. What seems to be certain is that instinct is often a functional and easy-to-use theoretical tool for the explanation of behaviors that are difficult to account for. As Bateson (1969) properly puts it in one of his metalogues, instinct hence ends up being a real explanatory principle:

> Daughter: *Daddy, what is an instinct?*
>
> Father: *An instinct, my dear, is an explanatory principle.*
>
> Daughter: *But what does it explain?*
>
> Father: *Anything – almost anything at all. Anything you want to explain.*
>
> Daughter: *Don't be silly. It doesn't explain gravity.*
>
> Father: *No, but this is because nobody wants instinct to explain gravity. If they did, it would explain it. We could simply say that the moon has an instinct whose strength varies inversely as the square of the distance...*
>
> Daughter: *But that's nonsense, Daddy.*
>
> Father: *Yes, surely. But it was you who mentioned 'instinct', not I.*
>
> Daughter: *All right – but then what does explain gravity?*
>
> Father: *Nothing, my dear, because gravity is an explanatory principle.*
>
> Daughter: *Oh.*
>
> (Bateson 1969: 11)

A better explanation of the concept than Bateson's can hardly be given. It is clever, witty, and ironic enough to emphasize how exaggerated a meaning we provide "instinct" with. Considering the number of actions, from the simplest to the most complex ones, that are described as instinctive, it shall be much easier to define the term in one of the following ways: "Supernatural power", "Invincible weapon", or—more humbly—"Ultimate solution to every problem". To which, it should also be added (a) where one can buy some instinct, (b) how much it costs, and (c) if it is legal.

The problem with instinct is its absolute conceptual flexibility. Once we have established that it is an impulse that drives animals (including humans) to perform a (limited) number of actions (and it can definitely be accepted that there are actions that we perform quite *before* having a mental representation of them), it becomes extremely easy to apply it to every case where a mental representation is not detectable (i.e., it is not verbalized by those who experience it).

Here lies the difference: if we produced our hypotheses in accordance with our observation, we would look at a beaver carefully cutting its pieces of wood in such a way that they all weigh the same (which is what actually happens: they *do* weigh the same. See the very interesting Richard 1967 on the topic), and we would say that, indeed, the beaver is carefully cutting his/her pieces of wood in such a way that they all weigh the same. With morganism, however, we are faced with a problem: our hypothesis now implies a superior psychic faculty. And this is very bad, if it is *possible* to interpret the same pattern as a result of an inferior faculty. We need an inferior one, that is, a faculty that does not require such an articulated mental activity (provided that estimating the weight of a piece of wood is "an articulated mental activity"). So, we "buy a dose of instinct": the beaver does not have the slightest clue of what is going on, but s/he has this supernatural power, which drives him/her to choose this piece of wood instead of that one, which is too heavy and too long.

Now, apart from the general theoretical remarks on instinct itself[53] (which hopefully are sufficiently illustrated, by now), the question is: was it *possible* to apply an inferior psychic faculty in this case? Because that is what Morgan says: we do not have to apply a superior faculty, when it is *possible* to apply an inferior one. Otherwise, one might say, we shall just leave the superior one, right? On the basis of which principle are we allowed to apply the notion of instinct on the beaver's weight estimations? Was it *possible*? Well, of course, everything is possible if we want it to be: but that is exactly how Bateson's explanatory principle works. It explains something simply because we deliberately designate it as explanatory. It does not explain it in a very scientific sense. So, what did Morgan mean by *possible*? Was it a *scientific* possibility or simply a *rhetorical* one?

---

[53]The discussion on instinct should not end with this note, as the abuse of the notion is, in my humble opinion, a real threat to scientific inquiry. Scholars should take a very critical approach towards those colleagues who so easily refer to this black box. In general, every application of the notion of instinct to complex multi-phased actions should be forbidden or accepted *only if* very convincingly and empirically motivated: it is all too easy to open the black box and put in processes and actions that we are not able to explain otherwise. Except that this is not science: it is much closer to religion. What is the difference between explaining a complex phenomenon with the existence of instinct and explaining it with the existence of God? And why not of any Marvel hero? It is necessary to be very firm on this also in order to protect the notion of instinct itself from massive trivialization. There *are* a number of human and non-human phenomena that originate at a pre-cognitive level and produce an actual physical action. But if one wants to make sense out of these instances, they first of all must be circumscribed: extemporaneous escape from danger, movements supporting physiological needs, innate competencies on immediate parental care… This is what should be called "instinct". Hunting strategies, elaborate parental care, establishment of social relations… if one calls those instances "instinct", there is no real need for any ethologist or zoosemiotician. A priest is enough.

2. What do we mean by 'superior' and 'inferior' psychic faculty? Do we mean 'more' or 'less' similar to the human psychic faculties? If yes, as one may bet, then we are typically facing another anthropocentric bias, that is, considering human intelligence as the universal (transpecific) parameter for cognition on this planet (or other planets, for that matter: do we not launch hypotheses on extraterrestrial life, on the basis of how more or less intelligent than us aliens can be?). If there is a transpecific paradigm to interpret psychic faculties that should probably be adaptive skills. The more adaptive the subject is, the more 'superior' his/her psychic faculties are. Gregory Bateson again makes an excellent point in this respect. In experimental contexts, he says (Bateson 1972: 368–370), the animal's intelligence is tested through four main points: (1) the animal "may or may not perceive a difference between the stimulus objects X and Y"; (2) the animal "may or may not perceive that this difference is a cue to behavior"; (3) the animal "may or may not perceive that the behavior in question has a good or bad effect upon reinforcement", that is, that doing the right action s/he will be granted with (usually) food; and (4) the animal "may or may not choose to do 'right', even after he knows which is right". As it can easily be deduced, the four points are set in order of priority, that is, 4 is possible if 3 is accomplished, 3 is possible if 2 is accomplished, etc. Therefore, step 4 reveals a superior psychic faculty than 3, and so forth. Now, are all these steps really tested in experimental contexts? Do trainers really go through the whole procedure? Bateson is pretty doubtful.

> Let me now consider for a moment the art of the animal trainer. From conversations with these highly skilled people – trainers of both dolphins and guide dogs – my impression is that the first requirement of a trainer is that he must be able to prevent the animal from exerting choice at the level of step 4. It must continually be made clear to the animal that when he knows what is the right thing to do in a given context, that is the only thing he can do, and no nonsense about it. In other words, it is a primary condition of circus success that the animal shall abrogate the use of certain higher levels of his intelligence. The art of the hypnotist is similar. (Bateson 1972: 369)

In other words: intelligence is tested by limiting intelligence. It reminds me of when I was a kid and adults would allow us to go playing outside on the condition that we would not sweat.

3. Any action, even the simplest one, never has a unique impact, and as a consequence can never be interpreted as monolith. A canon (or any other intellectual formulation) that promotes simplicity is in practice a canon that prefers to "lock" observations in such a way that they give the least possible disturbance.

Clever Hans was undoubtedly unable to perform mathematical calculations, so, to morganists, all we need to know is that this horse was not a good mathematician (which would have been a superior psychic faculty), but simply manipulated his observers (which appears to be an inferior psychic faculty). Period. And I cannot help wondering why we have to be so superficial with regards to that episode. And why we should evaluate the events only in a negative sense, as something (mathematical calculation) that *did not* happen. What we also have is a horse that, in front of dozens of observers, was always—or almost always—able to (a) detect an even microscopic facial or postural expression, (b) process these expressions mentally, and (c) understand which one, among the many, was to be interpreted as a message of approval.

And maybe there was even more to it than this. It was not only the ability to detect approval, it was also detecting the right degree of approval. Let us picture the scene: somebody asks Hans to indicate the square-root of 441 (21). The spectators gathered around the horse are very skeptical and determined not to applaud or congratulate unless the horse really stops his counting at 21. Thus, Hans starts to hit his hoof on the ground. After two or three hits there is already a quite interesting range of expressions among the observers: somebody is chatting, somebody is laughing, somebody is betting that the horse will not even reach 10, and so on. Everything is accompanied by some non-verbal patterns that vary according to the related emotional state, but also to the subject involved (let us just think of how many ways to laugh/smile/giggle exist). Hans understands that all these expressions are not relevant, and keeps on hitting his hoof. After some ten hits, somebody already goes like "Well… not bad!", and in general there is an increasing attention towards the scene. In this case, too, Hans understands the irrelevance of such expressions, and proceeds with his counting. Here we are at 18–19 hits: people grow excited and surprised: "hey, he's doing it! He's really making it!". These are all expressions that one might easily mistake for a "definitive" approval, the one that Hans is looking for and that will mark the end of his efforts. But no, the horse does not stop yet: he understands that these are only forms of "quasi-approval". He reaches 21, and then, only then, he understands that the type of expression he was looking for is now printed in almost everybody's face: it is the *final* approval, the one that suggests he should stop, because people are happy with his performance.

Now, a hypnotist or a psychologist could hardly do better: this is for sure. Still, to morganists, all this seems not to have any importance: what matters is that Hans was not an equine Einstein. To be an equine Freud is not enough. Why?

4. The most direct consequence of morganism (the fear of anthropomorphism), too, presents several obscure points. What is really anthropomorphic, given that many characteristics are human but not only human? Is anthropomorphism really to be condemned in toto? And most of all, could the fear of anthropomorphism be more dangerous for scientific research than anthropomorphism itself? In the last few decades, several scholars in animal studies developed a less hostile attitude towards anthropomorphism. They for instance realized that promoting a certain empathy between humans and other animals does not necessarily constitute an anthropomorphic mistake, and that neglecting it might be even more misleading. An example that is really worth mentioning, called "critical anthropomorphism", is analyzed by Luisella Battaglia, in her essay *Etica e diritti degli animali*:

> Critical anthropomorphism aims to use human experience in a critical manner, in order to recognize emotional manifestations, putting into relation our most immediate subjective intuitions with comparable notions and data provided by neurophysiology, ethology, zoology etc. This way, empathy, typical of classical anthropomorphism, is integrated with the most recent scientific research on animal life and behavior. In support of such an approach, we could say that if human-animal similarities are accepted for scientific experimentation purposes, they should also be accepted in the field of emotional sensibility. In any case, any doubt about emotional sensibility should be gauged so as to benefit the weakest subject. In particular, the presupposition of similarity, when there is no clear counterproof, should be interpreted in favor of the animals. (translated from Battaglia 1997: 123–124)

In other words, if, as often happens, scientific research involving non-human animals—either aiming to study animals themselves, or aiming to study humans *through* other animals, as in medical research—is based on presuppositions of similarity between humans and other animals, then it shall be fair to apply the same principle on a more consistent basis.

> It is not fair […] to maintain that there is a resemblance between humans and animals when we use [the latter] in medical laboratories in order to establish our rights over them, and – at the same time – to maintain [that animals are different] as moral subjects, in order to avoid our duties. (translated from Battaglia 1997: 124)

Battaglia's main point here is that philosophical speculation on animal behavior (as on other issues, as well) should coordinate scientific data and ethical reflections, in such a way that the latter support the former, when these are not able to provide an exhaustive response. When science is not able to give a 100 % reliable answer on a given behavioral phenomenon, than ethics should be allowed to have the final word. And ethics, as the conceptual basis for modern jurisprudence, is based on the principle that the weaker subjects, those who are not able to verbalize what they are doing and what they mean by that, should be favored and protected. Such is the case with categories of subjects like infants or mentally-disadvantaged people. The problem that a small child is not yet able to understand the question "would you like to be hurt?" is solved (scientifically and ethically) with the assumption that he would probably say "no", not "yes".

Going back to the specifically scientific aspect, Battaglia's conclusion is that all the generally emotional manifestations previously considered dangerous for scientific research (empathy, most of all, as the generator of the CHE par excellence), if handled with care, become useful and plausible clues for the whole theoretical apparatus. This is because: (1) by definition, empathy helps comprehension; hence it is easier to interpret a given phenomenon if we identify ourselves with it to some extent; and because (2) animals are not only subjects of human scientific research: they are also the category humans belong to. Humans are not *more or less* similar to other animals: they *are* animals. To ignore this fact means to ignore a crucial part of the story.

To add a specific semiotic note to it, the concept of critical anthropomorphism, and empathy in particular, promote nothing other than Peircean abduction, as a valuable method for scientific inquiry.

With these four main contradictions in mind, I believe we have solid ground to work on an alternative (numanistic?) concept to morganism. We of course need a concept that is not exclusively in opposition to something, but also in favor of something else. When I first suggested this idea, in a seminar that took place in Estonia in 2004, my colleague and friend Kalevi Kull suggested that I should call it "Martinelli's Canon". As tempting as this may be for my ego, I suggest we do not name it yet, and we rather wait for a more complete articulation of these ideas, and—hopefully—some feedback.

At any rate, if we formulate an *alternative* canon, then we should be more or less define its basic rule as follows: *In no case should actions or behaviors be interpreted as the result of an inferior psychic faculty, when it is possible to interpret them as a result of a superior faculty*. Comments:

1. Unlike morganism, the idea here is to *approach the problem of inferior and superior psychic faculties in terms of abductive, not rhetorical, possibility*. A behavioral phenomenon X (e.g., carrying pieces of wood that weigh all the same) is observed in non-human animals (e.g., a beaver); X can be explained with hypothesis A (the beaver is mentally estimating the weight of the pieces) or B (the beaver is driven by instinct); X resembles A on the basis of perceived similarity with A_H (i.e., A as performed by human beings); hence there is a reason to pursue A. Pursuing A does not mean automatically *stating* A as the final word on the matter. It means that our *research hypothesis* is now A, and the next step of our inquiry should be *oriented* towards A. If that proves to be a failure, that is if pursuing A is not *possible*, then further hypotheses shall be considered. In this sense,
2. Unlike the deceiving claims of morganism, our alternative canon *does* aim at *economic and ergonomic research*. Pursuing A is more economic and ergonomic than pursuing B, because A already provides clues and research tools, therefore a path to follow has already been indicated.
3. Also, we should *refuse the criteria for establishing differences between psychic faculties in the way morganism does*. Such criteria, rather than anthropocentric, should focus on pluralistic notions such as that of Umwelt, adaptation, free will, etc. Interpretive hypotheses that rely on concepts like instinct are by principle denying such pluralism.
4. Unlike morganism, we should make a genuine effort to *treat any behavioral phenomenon as complex and multi-layered and refuse in principle reductive one-sided interpretations*. An instance like the one involving the horse Hans represents a paradigmatic example of a multi-faceted semiotically-rich subject for scientific inquiry, which scientists reduced to a quasi-joke by refusing to approach the episode in all its aspects. Once it had been established that the horse was not capable of mathematic computing, they ignored his still-astounding ability to interpret body-language, therefore missing a good chance to deepen the knowledge of horse communication in relation to their Umwelt.

5. We should supports notions like *critical anthropomorphism* (Battaglia 1997: 123–124), *and therefore reject reductive hypotheses on an ethical basis, as well.* If the interpretation of the behavioral phenomenon X is complicated by the impossibility of making a totally reliable scientific choice between the hypotheses A and B, then—in addition to the motivation provided in point 1 of the present list—another reason to pursue A is the ethical one: "the presupposition of similarity, when there is no clear counterproof, should be interpreted in favor of the animals" (translated from Battaglia 1997: 124).
6. On a more strictly scientific level, we should *promote methodological empathy and*, more generally, *support a more balanced and less radical use of anthropomorphism.* Anthropomorphism should be avoided when it applies human species-specific predicates to non-human species and when it produces distorted perceptions of the reality analyzed. It should not be avoided, but in fact encouraged, when, through empathy, it helps comprehend the given phenomenon, when it favors biocentric approaches to non-human species, and when it is employed for didactic purposes, in order to facilitate attitudes of inter-specific acceptance and compassion.

Chapter 4
Appendix: A Short (Pre)History of Numanities

4.1 Introduction

This "Appendix" document, in an ideal world, should have been the third and final case study of the book. It is meant to close the circle of this "official launch" of the Numanities platform, and it is a meta-case because it describes the genesis and development of Numanities at an institutional, organizational, conceptual and operative level, as opposed to the purely scholarly one. Doing so was one of my last decisions, while forming the contents for this book. I was indeed not fully convinced that this would have been a very good idea within the context of an academic monograph, which should rather aim to present data and relative comments. For this reason, I came to the compromise of including it as chapter, fearing that some reader would have found it "not academic enough". Plus, I am not helping the situation by writing this part in a much more informal, occasionally "literary" style. Nevertheless, I still would like to call this section a "case study", regardless of its location in the book.

Two factors convinced me of the justness of this choice. First and foremost, the second article of the manifesto: *Re-establish the dialogue between humanities and institutions, at various levels.* If we really want this dialogue, and I suggest that we do, then—besides the mentioned need to learn the "rules of the game" of modern society—we should also keep track of the institutional aspects of what we do. An "institutional history" of a given research concept (or project, or paradigm…) should not be disregarded, as this is one of the ways for us not to forget the social, economic and political context in which that concept is generated. Even the most abstract of philosophical concepts originates from a combination of elements that are not exclusively internal to philosophy and abstractness.

That "architecture of knowledge" we spoke about in the manifesto would not be practicable if we did not accept the many implications of this metaphor of a field like "architecture": there are foundations to a building, there are "materials" in its

D. Martinelli, *Arts and Humanities in Progress*, Numanities - Arts and Humanities in Progress 1, DOI 10.1007/978-3-319-45553-2_4

construction, and they all must come from somewhere that is *outside* architecture, be they the wood becoming a door, the stone becoming a tile, or the metal becoming a pipe.

Plus, shall we say it all? We cannot deny that plenty of so-called "scientific" choices in a given scholarly activity relate in fact to institutional strategies much more than to individual inclinations. Nowadays more than ever. And, within reasonable limits, there is nothing wrong with that: being part of an academic institution means also being part of a team (a "family", if we are lucky). Not everything can be decided in total autonomy: a researcher has the target of his/her personal growth, but also the institution's growth.

The second important factor that convinced me to compile this rather peculiar case study was the very recent appearance of Eero Tarasti's book *Sein und Schein* (2015). Prof. Tarasti, who I have mentioned on more than one occasion here, was my supervisor at the times of my doctoral thesis, a defining example of mentorship for me, and finally I am also happy and honored to enjoy his and his wife Eila's close friendship. In this recent book of his, Tarasti, who, since the 1980s, has been a pivotal figure in most of the important institutions and projects in world semiotics, wrote in that book a whole chapter entitled *Can Semiotics be organized? Observations over a 40-year period* (starting from page 431 of the book until the end), in which he does exactly what I plan to do now. That is: taking the "institutional/organizational history" of semiotics as an opportunity to inform us not only of events and projects, but mostly to emphasize the bond between actual scientific developments and, say, the appearance of this or that scholar, the organization of this or that seminar, the start of this or that project.

In sum: this is nothing other than the old recommendation that our school teachers used to give us, whenever the students of my generation would approach the study of an artist, a philosopher or the likes. That is: we cannot understand the contents of someone's ideas and actions if we do not study their life. Is it possible (or even realistic) to understand Kafka's work by ignoring his life? Would Henri Tajfel have ever written about ingroups and outgroups had he not gone through the tragic experience of the concentration camps? Would Dante have ever written the *Divina Commedia* in that particular way (or at all) had he been more successful with Beatrice?

In this (meta) case study, thus, my intention is to tell the pre-history and early history of Numanities, starting from their establishment within the International Semiotics Institute at Kaunas University of Technology (including *how* this institute was established at this university, and *why* it generated something like Numanities), and continuing with the first research projects, the first events, and a bit of future plans as well.

A last remark is perhaps called for. I fear that the readers may perceive that this appendix is, in some points, more of a show-off or a commercial spot of the International Semiotics Institute, rather than a real description of events and activities. This is possibly because I made the decision not to sound too diplomatic about our achievements, but simply telling them like they are, nor did I hide the fact that we felt very proud of them. To mention one: as an Italian, I know very well how difficult and dramatic is the struggle of young researchers to find a decent occupation in the academic world, and I also know how the older generation is for the 90 % selfish and insensitive to it (or too busy hiring their relatives and partners). Now, to be able, as we did, to hire a few young, talented scholars in the institute, is something I am shamelessly happy about and proud of, and I have no intention of hiding these feelings, particularly in front of that 90 % of senior colleagues.

At the same time, I humbly ask these readers to pay attention to the many occasions in which I emphasize our various problems, limitations and also failures: hopefully, they will see that they are not few and they are not small. I conclude this appendix with a SWOT analysis of the institute: solomonically (but also honestly), there are 5 strengths, 5 weaknesses, 5 opportunities and 5 threats. Seems fair to me.

4.2 The Finnish Years of the International Semiotics Institute

Numanities, as a concept and as set of research activities, are—as I repeatedly mentioned—a platform established and implemented at the International Semiotics Institute, in Kaunas, Lithuania, within the premises of Kaunas University of Technology, an institution that stems from the oldest academic institution established in that town in 1922 (incidentally, the very first building, erected back then, is also the one that, beautifully renovated, hosts now the Faculty of Social Sciences, Arts and Humanities, where the institute has its headquarters). Why would a concept like this emerge in this particular context? I shall first offer a historical overview of this process, and then—in the next paragraph—I will also mention some strictly scientific reflection.

This paragraph—describing the establishment of the International Semiotics Institute *before* it moved to Kaunas—will draw heavily from the mentioned Tarasti (2015: 436–448), to whom I can only repeat my gratitude. Then, later, I shall take from where Tarasti's book leaves: January 2014, the date of the opening of the institute in Kaunas.

Tarasti locates in 1986 the "big bang" of what will eventually become the institute, when Pentti Rossi, director of a summer university in Imatra (a small Finnish town close to the Russian border), and Henri Broms, vice-president of the Semiotic Society of Finland, decided to organize a symposium in semiotics at the Imatra Cultural Center, a new white building by the river Vuoksi, close to the rapids that had once made Imatra a famous attraction for Russian aristocracy. Thomas Sebeok, one of the then-living most important semioticians, was also invited, and along came also a bunch of famous musical semioticians, such as Costin Miereanu, François Delalande, Gino Stefani, Marcello Castellana, Luiz Heitor Correa Azevedo and Tarasti himself. These presences, Sebeok on the one hand and the musical semiotic group on the other, would become two important seeds for the International Semiotics Institute: the bond with the Toronto semioticians (via Sebeok), and the prominent role always played by musical semiotics at the ISI events (in particular, the launch of the satellite project of "Musical Signification" which has generated, as of 2015, 13 international congresses around European towns and 12 international doctoral and postdoctoral schools of musical semiotics at Helsinki University and Imatra).

A second edition of this semiotics symposium in 1987 also saw the appearance of prominent members of the Tartu-Moscow semiotic school, including Alexander Piatigorski, Boris Oguibenin and Anne Shukman, creating another important bond. In the same year, also the Nordic Association of Semiotics started its activities.

The crucial year, however, turned out to be 1988, when the International Semiotics Institute was actually established as an association, with a key-role played by Prof. Paul Bouissac in the process:

> The Canadian semiotician Paul Bouissac, who in Toronto was running his famous summer schools for structural and semiotic studies, which always lasted one full month, had got the idea to establish an institute to supervise the education of young semioticians and their teaching globally. There had to be a place where one could know everything that happened in semiotics in the whole world; a database from which students would learn where to go in order to specialize in diverse fields of semiotics. He had built a collegium of 44 scholars behind this enterprise, but it was certainly Sebeok's initiative which brought the founding meeting to Imatra. The candidates for the presidency of the ISI were Paul Bouissac and Roland Posner, but in the end I [Eero Tarasti] was elected. (Tarasti 2015: 429–430)

While the goal of establishing a database for semiotic activities was never really achieved, ISI became quickly active in organizing international symposia and congresses, placing Imatra on the map of the semiotic world. In particular, ISI inherited from the Toronto Semiotic Circle (once the latter was no more able to carry it out) the Summer School of Semiotics, and that soon became the flagship event of the institute, taking place every June (in Fig. 4.1, two of the most successful sessions of the school). Initially a one-week event, the School added another week specifically for Finnish students and scholars in February (becoming thus a Winter School as well).

The growth of ISI was exponential year after year, encouraged by the semiotic community's increasing enthusiasm and by successful fundraising activities (partly because of the merit of the institute's team, and partly due to the many funding resources that Finland could offer in those days). In 1992, Tarasti launched a book series called Acta Semiotica Fennica (still active nowadays). By the late 20th and early 21st century, the participation in the Summer School reached an average of 100 presentations per year (including myself, for the record: my first participation in the Summer School was dated 1999, and since then I must have missed not more than two or three editions. Moreover, starting from 2003, I started being involved in the organizing and/or scientific committees of the various events organized by the institute).

By 2007, with Tarasti also having become the president of the International Association for Semiotic Studies, IASS-AIS, the world's largest organization in semiotics, ISI was also ready to host the world congress of semiotics (in Fig. 4.2, two publicity items for that event). That year, nearly 1000 semioticians convened

Fig. 4.1 *From left to right*, two "classic" sessions within the Summer School of Semiotics in Imatra, Finland: biosemiotics (2000 edition in the picture) and musical semiotics (1999 edition in the picture)

Fig. 4.2 *From left to right*, the poster and the homepage of the 9th World Congress of Semiotics, hosted by the International Semiotics Institute in 2007

from all over the world in Imatra and also Helsinki (whose University co-hosted the event). As I shall later mention, ISI—in its Lithuanian incarnation—will have the honor to host the event again in its 2017 edition.

Additional achievements of ISI in that period were—among others—the establishment of a Network University of Semiotics, a 4-year project led by Tarasti and Harri Veivo (who got a position as professor within that project), with a little help from myself and a talented German semiotician named Guido Ipsen (both nominated guest-professors); and—few years later—the European PhD program "SemKnow", in collaboration with the universities of Lapland, Tartu, Torino and Sofia (the program is still ongoing).

These events, all happening in the second half of the 2000s, represented the peak of ISI's Finnish phase. After that, as it often happens to anything reaching "a peak", things started to crack a bit. Encouraged by the success of the Network, the University of Helsinki finally agreed to establish a Master in Semiotics, and therefore opened a competition for a full professorship. It was, finally, the acknowledgment of semiotics as a discipline of its own, and not as an chapter of philosophy (or linguistics, or art research, or else), as it had been until then in Helsinki.

I have never understood exactly what happened, why, and because of whom, but a euphemistically-surprising decision followed as a local and international jury unanimously chose the least likely of the candidates for the job: a Finnish philosopher, with little professional interest in semiotics. It was an official "invalidation" of semiotics as an autonomous discipline, and a glamorous step backwards taken by Helsinki University: sorry guys, after all semiotics has to remain a chapter of philosophy. It goes without saying, the Master's program ended up a fiasco, and did not even complete the first cycle before attestation, as Helsinki University closed it one year earlier. As much as one would like not to engage in conspiracy theory, this accident looked like a carefully-planned "perfect murder" of Finnish semiotics.

Along with that, due also to the financial crisis and the endless stream of reforms in academic institutions, funds began to decline. Tarasti and his team made an enormous effort to keep ISI alive, succeeding nearly every year to still arrange the Winter and the Summer Schools, plus other events and the publications of Acta Semiotica Fennica. But, admittedly, things were not as they used to be, also in terms of participation in the events (by then, not anymore touching three figures). Looking back, it is no less than amazing how Tarasti in particular (but also some of his collaborators) managed to keep things from collapsing and still organize high-quality meetings. One of the chief characteristics of the Imatra schools was its cosy atmosphere. Participants would always and easily befriend each other, and establish a sense of camaraderie that would be extended to other events as well: you would go to some other congress, and if you happened to meet a colleague "who also was in Imatra", that, for sure, would be your company for those days.

4.3 Bringing the Institute to Kaunas

Around 2012, the history of the institute started to intersect very closely with my own. During that year, indeed, after almost 13 years spent in Finland, first as a doctoral student, then as a researcher and finally as adjunct professor in semiotics and musicology, my family and I moved to Vilnius, and I started working as full professor at Kaunas University of Technology (KTU), at the then Faculty of Humanities. They were turbulent days for this place as well. On my arrival, the faculty had no stable dean, with the position filled pro-tempore by Audronė Daubarienė, a very gentle person whom I instantly befriended and who, I would soon discover, was one of the real training forces of the whole faculty. I mention her name as a Chekhov's rifle: we shall read more of her later.

The search for the new dean was related to another important novelty. The faculty was scheduled to merge with Social Sciences, departments were going to be re-arranged, resources optimized (a delicate expression for "firing people"), and most of the leadership had anyway to change, as they all had concluded two 5-year terms (an occurrence that most leaders did not like, but in Lithuania—as in many other civilized countries—a leader cannot occupy his/her position for more than two terms). Inevitably, the new dean, whoever he or she would be, had to be in the eye of the hurricane. It was thus not too surprising to see that, at the beginning of 2013, the successor to Daubarienė, a well-known and talented writer named Dalia Staponkutė, had the most difficult time in managing the faculty. In between the rock of the reforms and the hard place of some department leader's anger, Staponkutė decided to resign after few months. It was, possibly, the most challenging period for anybody in the faculty, veterans and newcomers like myself, conservatives and innovators, heroes and villains (roles that were of course switching, depending on the point of view). And it was also the period when I witnessed, unmistakably, how academic environments can be just as bitchy and cynical as any other place: an additional task for the next leaders had to be also the reconstruction of a sense of community and cooperation, which—at this point—was buried into an "every man for himself" hole of resentfulness, intrigues and individualisms.

A little while before the arrival of Staponkutė, almost out of the blue, I had received an email from Eero Tarasti, with a stunning announcement. He had decided to close the International Semiotics Institute in Finland: he was close to retirement, the fundraising process had become problematic, and semiotics was clearly agonizing in Finland—the next edition of the Summer School, in June 2013, was going to be the last one. And, well, this was not even the most astonishing part of the letter. How was I?—he asked—How stable was my situation at KTU, after these first few months? Would I be interested in taking over the institute and bringing it to Kaunas?

Gulp!

The proposal came so surprisingly that I almost failed to realize what a huge honor it implied. Eero, who throughout his career had always been supportive of his students and had generated plenty of disciples and potential "heirs" all over the world (literally), was singling me out from this crowd of prestigious colleagues as the one who could take over the most important and treasured result of his work: the institute.

Instinctively, of course, I was all the way inclined to accept this tremendously flattering invitation. In practice, I had to do a lot of talking with my "bosses" at KTU: ISI had been an association all this time, with the pros and cons of this status. Among the former, there was a relative but solid freedom of initiative. Now, were the plans to materialize, ISI was going to become an actual unit within the faculty, a research institute with the same status of a department. That would have meant certainly more institutional prestige and many other pros, but certainly freedom of initiative was not among those (and rightly so, to an extent). To be a research institute in the full sense of the term means to conform to the general strategies of the faculty (which in turn conforms to the general strategies of the whole university), it means to carefully plan your budget, it means to go through regulated procedures for nearly every administrative step you want to take, and so forth.

Most of all, my bosses wanted guarantees, and I wanted some as well, and they were not the same guarantees. I wanted to retain as much as possible the spirit of the Finnish ISI (starting from the above-mentioned cosiness), my bosses wanted a functional unit that would not be a burden for KTU, particularly on the financial side. I wanted to introduce creativity and lateral thinking in most of our activities, my bosses wanted to align the institute to the general branding of KTU (that, too, in a stage of reformation, in those days). I prepared a research plan—my bosses asked me for a business plan, instead. My main interlocutor, at this stage, was the vice-rector for research, Asta Pundzienė, a person I immediately came to admire for her intelligence and decision-oriented attitude. I cannot speak for her, but I think we both had an understanding of the different roles we had to play in this situation, and we both respected those roles. There was never any hint of animosity in whatever compromise we had to reach, up to the point that I do not even call them compromises, but simply cooperative planning.

As Dalia Staponkutė came onto the scene, I have to say that she gave me enormous support in the transition of the institute from Imatra to Kaunas. Already very acquainted with semiotics, she never hesitated a moment to show her interest in this project, and she used the faculty budget to both invite Eero in Kaunas, for some preliminary agreement, and send me to Finland for the final Imatra Summer School in June 2013. Audronė had unfortunately left her pro-tempore position before my communication with Eero, and for a few weeks the faculty remained without a leader: I am sure she would have been equally supportive, had she been in the position to be so.

So came June 2013, one of the most peculiar experiences in my career. The first months at KTU had already taught me a lot in organizational terms. During my 13 years in Finland, first as a doctoral student, then as a post-doc and finally as adjunct professor, I had been happily protected from administration: with very occasional exceptions, all I was always asked to do was research and teaching. A privileged condition, when I think about it. At KTU the change was radical, also because my position had become more important and filled with more responsibilities: now we were talking research and teaching, *plus* administration, organization, strategic planning, business planning, SWOT analyses every second week, team-building sessions, and dozens of other tasks. Although I have been complaining several times about the overload of such activities (even in this book), I still have to admit (and I hope none of my bosses will read these lines, otherwise they will be used to embarrass me at the first possible occasion) that I learned a lot from this, and I became a much better professional in many things I do.

So, here I was, heading to Imatra with a very complete and rational plan for the new ISI in Kaunas, confident that my Finnish colleagues and friends would be very happy to learn that their beloved institute was after all going to survive. The agreement with Eero was that we would have announced this new decision only during a specific plenary session on the future of ISI. First, he would have taken the floor announcing the end of the Finnish years, and then I would have stepped in announcing the beginning of the Lithuanian era.

The session started: Eero used all his rhetorical skills (which, believe me, are mighty) to express his sadness about the way events turned out, how Finland was now detaching itself from semiotics, how it had become difficult to convince institutions of the importance of this field, and how—finally—this was going to be the last Imatra school and the end of ISI. Faces all around were either funeral-like or straight away weeping. I, on the contrary, was more and more anxious: "When is he going to say it? When is he going to say it? When is he going to say that it is NOT over, that life goes on, that there is future for this institute???"

Nothing. He did not say it. To this day, I do not know if he simply forgot, or if, after all, he meant that that was a kind of ending anyway. To make matters worse, Prof. Pirjo Kukkonen, a renown translation scholar and leader of a symposium in Semiotics of Translation that took place every year in Imatra within the school, took the word to say how much she was saddened and upset by this state of things, and how much she was grateful to Eero for having run the institute so greatly until its end.

Until its end.

Oh, and by the way, it is Dario Martinelli's turn to speak, now.

I do not remember what exact words did Eero emply to announce me, but my mind distinctively heard "There you go, Dario: follow that!!!". With my tail between the legs, I took the floor, knowing that what *I thought* Eero would do—that is, cheering up the audience with hope and future plans, that I only was supposed to explain and describe—was now up to me. From my position, in front of the audience, I could see even more clearly the level of sadness and hopelessness of every single face (and we were about fifty, that afternoon). My plan for the beginning of my talk was now entirely screwed up, and I had to come up with something else in those very seconds between standing up from my chair and reaching the speaker's desk.

I do not think I have many things in common with Steve Jobs, but there is one where we are exactly alike: we both think that whatever happens in life there is either a Beatles' or a Dylan's song that can capture that situation in poetry. May

Lennon-McCartney and "Hey Jude" forever be blessed, because those lyrics came to rescue me in those very seconds. I inaugurated my speech with "don't make it bad, take a sad song and make it better, remember to let her into your heart, then you can start to make it better". In retrospective, quoting "Blackbird" may have even been more appropriate (with that idea of taking "these broken wings and learn to fly"), but "Hey Jude" did its job well enough. I told my audience that we did not have to "make it bad", or "carry the world upon our shoulders", but rather that it was up to us ("and don't you know that it's just you!") now to make the song better, by letting her into our hearts (which meant basically that we had to preserve the same spirit, the same feelings, and simply move them somewhere else, without losing them).

In truth, it is not like I cheered up just everybody (my colleagues from Finland especially remained sad throughout the whole evening, no matter what I could say), but certainly I was now in the position to continue the presentation with my original, carefully-detailed, plan. Very encouraging feedback came particularly from my Indian colleague and friend Seema Khanwalkar, and from two young talented students called Majid Bahboudi and Alin Olteanu. Majid, in particular, would become a crucial figure in the early development of this new stage, by volunteering (I cannot thank him enough for this) to collaborate with me on the visual re-branding of the institute. What a visitor can see on our website, for instance, is still primarily Majid's work (Fig. 4.3).

Fig. 4.3 The official announcement of the new "Lithuanian" ISI, during the last Summer School of Semiotics, June 2013

4.4 Right Before the Opening of the New ISI

The official opening of the new ISI occurred in January 2014, but some anticipatory work had already been done in the autumn of 2013. At this point, we had become the Faculty of Social Sciences, Arts and Humanities. Dalia Staponkutė had left in the summer of 2013, and, after a short period in which the dean's office was occupied by a business scholar named Monika Petraitė, the "scepter" went to Jurgita Šiugždinienė, a social scientist with a strong background in public administration and experience in government institutions. The situation had become much more stable in the faculty: the transition had occurred, some leadership had changed, and reforms were met with a bit less resistance. Having said that, the difference Jurgita managed to make in the process was enormous. She rejuvenated the whole team, she took courageous decisions, she created a tangible sense of community in all of us, she had a solid, and equally gentle, sense of leadership, and she strengthened both the "form" and the "contents" of the faculty. For the first time, after the beginning of the reforms, we all felt like we were going somewhere.

It was, thus, a much more welcoming atmosphere for the ISI to develop properly in the new environment. It was time for me to build a team. I knew all along that whatever step I would take in KTU, I would have loved to have Audronė Daubarienė with me. Audra (as we all call her) is a rare combination of efficiency, intelligence and kindness, she is pragmatic without being a "bureaucrat", and she is also a lot of fun to work with. As we were talking about a head of administration for the institute (therefore the collaborator I would have to deal with on a daily basis), I could not think of a better choice. I spoke to Asta Pundzienė, and she created an administrator position for the institute, which was filled by Audra. The nucleus was created. Asta was also very gracious in giving me a researcher position to fill in any way I would like. It may be too long to explain, but the system in Lithuania is such that work is calculated by "staff units" (*etatas*, in Lithuanian). When one has 1 staff unit, it means s/he has full-time employment. So, s/he can have less (0.2, 0.5… any fraction of 1, basically), and in principle more as well (the law allows up to 1.5 etatas), if s/he can handle the time and the workload.

It must be specified that Lithuanian academic institutions are hardly granting anymore "staff units" in research. They give them for "teaching", and with those a scholar has to find time for research too. But to have a position *only* as researcher is a luxury given in very few cases. The fact that I could now distribute 1 etatas just for research was crucial in developing the institute. I did not want (and still do not, at the time I am writing) to get involved in pedagogical tasks (which I of course perform for other departments in the faculty), however great that would be, because a study program is a very demanding activity for a department/institute, and if you are a small unit like we were (and still are), it can absorb all your energies, time and budget. To be able to preserve the "research institute" identity for the ISI was very important for whatever activity we were planning. Having been given a kick-start with staff units for administration and research, the agreement was, and still is, that any expansion of the research personnel would have been up to our capacity to generate positions via research projects. Fair enough.

As already mentioned, some preparation work had been done before January. We had two important tasks to attend to before the official opening, plus another one that was arranged quite spontaneously. At this stage, it was only Audra and I who managed the yet-unborn institute, but there are at least two more people who helped out significantly in this early stage: one, who I have already acknowledged, was Majid Behboudi. Another one was Saulė Petronienė, then head of the Foreign Language Center (and now head of the Department of Modern Languages and Intercultural Communication), another very nice and smart colleague. It is of course impossible to mention all the people who, at some point, gave us a hand in growing up, but—in the course of this paragraph—I shall do my best to mention as many as possible, first as a token of gratitude, and second in the attempt to get this "history" as accurate as possible.

The first important task was the continuation of the Summer School: every colleague who has ever organized a congress (especially an international one that gathers around one hundred scholars from everywhere) knows that as soon as you finish one year's edition of that event, it is already time to think about the next one. In that sense, we were already late. I will mention all this preparation, and most of all its contents, in the paragraphs "The ICoN event".

The second important task was the transferring of some tangible heritage from Imatra to Kaunas. Books, primarily. During its Finnish years, ISI had produced a significant body of publications, via the Acta Semiotica Fennica series, which included congress proceedings, PhD theses and monographs. But also, and perhaps more significantly, one of the biggest semiotic libraries of the world had been assembled during all these years. In collaboration with the Imatra City Library (which invested a lot of money in this enterprise), ISI had purchased books and collected all sorts of donations from other academic institutions and individual scholars (a nice tradition in the Summer School was that whoever had a new book out, would donate at least a copy to ISI). Dozens of boxes full of books had to be moved from Imatra to Kaunas. The Imatra City Library was incredibly generous to let us have the majority of the books they had themselves bought, totally *gratis*. We still have to find a nice way to express our gratitude properly, but we will (Fig. 4.4).

A van was sent to Imatra to collect all these boxes. As they came to Kaunas and we started opening them, I had a bit of an emotional moment. Here was a room full of boxes that were the result of 25 years of loving labor. If you work in the academic world, books are the ultimate currency of your efforts: you work with books, you aim to write books, you teach through books, you give and (are very happy to) receive books as gifts… for the first time, I truly understood why Eero and all the Finnish colleagues were so melancholic on that particular occasion in Imatra. I bored Audra to death, that day we opened the boxes, by basically attaching a little story to every book that appeared before our eyes: "Oh! This I had as

Fig. 4.4 Eero Tarasti and his assistants pack the books (destination Kaunas) accumulated and printed in 25 years of activity of ISI in Finland. A historical, but certainly bittersweet, moment

textbook in one course!", "This you should read: it relates to your field", "This I made the cover of", "This is the PhD thesis of my office-mate in Helsinki"…

If ever I needed one, this was the ultimate proof of how much history was behind ISI, and how we had to do all we could to preserve it as an "affective history", not just an "institutional" one: *remember to let her into your heart...*

Then, as I said, a third task arose quite spontaneously. On December 5, 2013, the world cried the passing of one of the greatest individuals who inhabited it: Nelson Mandela. We soon thought that the man who had said that "Education is the most powerful weapon which you can use to change the world" had to be honored with an educational event. Literally in one week, we quickly arranged a one-day seminar in which we would present Mandela's life and work, and where we would screen some audiovisual material about him (we chose the documentary "Amandla! A Revolution in Four Part-Harmony" and Clint Eastwood's movie "Invictus", for the record). As the event was a non-stop 7-h-long "marathon", we chose the title "Mandelathon". It became the very first event organized by the new ISI, and took place in the main lecture hall of the old Faculty of Humanities in Kaunas.

4.5 The New Era Officially Begins

As Mandelathon closed 2013 on a meaningful note, 2014 started with the formal opening of the ISI (it started also with a very difficult period in my private life, but that is a whole different story). We purposively decided not to do any pompous inauguration, wanting to keep all the festivities for the first edition of the new congress, now scheduled for early June.

Partly because of what we had inherited from Finland, partly because of the novelties we wanted to introduce, and partly, finally, because of how things naturally developed in 2013, we could already draft some action plan for the new ISI.

Obviously, we intended to perform a smooth transition from Finland to Lithuania, trying to retain some of the most important features of the institute's history, and maintaining strong connections with Finland, through its key figures, of course, but also trying to persuade most of the many Finnish scholars who had been attending the events in Imatra, to "follow us" to Kaunas. We always loved that perfectly symmetric logo, which—since the early years—had retained its modernity, so there was no need to change that either. Needless to say, we created an "honorary director" position for Eero Tarasti, and a branch office in Helsinki, where Eero can continue some ISI-related activities, like the Acta Semiotica Fennica series, which is still fully operative.

At the same time, we planned a number of important innovations, to implement immediately or in the longer term, and these originated both from the change of personnel, whom of course would bring their own research inclinations and personal style (I shall soon return on the staff as such) and from the institutional status: the ISI was now part of a humanistic faculty within a technological university. That, alone, sounded (and is) like an interdisciplinary statement. In addition to this, a certain crisis of appeal of semiotics (and humanities in general), and the increasing difficulty among young researchers to find appropriate outlets for their skills and talents, could not be ignored.

For all these reasons, it became soon clear that the new ISI had to take a clear, ethical and scientific, stand within the academic community, as an institute that meant to:

1. Be *in* and *about* society. Priority had to be put to those research activities that discuss problems of current concern, that have an impact on society and that are ultimately not detached by social, political and cultural issues. The significance of Mandelathon as our first event was now even more evident.
2. Take particular care in understanding progress and technology, with the purpose of assessing them for their ability to provide quality and dignity of life (including scientific research as such).
3. Adopt a new approach that would be closer to the younger generations, not only in terms of contents and opportunities, but also in terms of image and communication. Sure enough, we soon opened a Twitter and a Facebook account, and we also launched some "viral" initiatives tailored for academics, like our shelfie campaign (#isi_shelfie), our symbolic protest against that redundant

celebration of the ego called "selfie", which instead celebrates culture, reading and knowledge, by taking pictures of our own bookshelves (for the entire picture collection, please visit the following link: https://www.facebook.com/media/set/?set=a.482217191948002.1073741830.228656963970694&type=3—and perhaps send your own shelfie!).

4. Carry out research in a truly interdisciplinary sense. If there is one strength that semiotics has displayed in its history—we thought—that is certainly its intrinsic nature as "methodology" (rather than "theory") applicable to the most diverse fields. At the same time, semioticians had often failed in opening up to other disciplines, or—when they did—it was often with the purpose of stating that, after all, those other disciplines were "inherently semiotic", thus closing the door again.

To encapsulate the scientific and ethical mission of the new ISI we thought of two zoological acronyms: CATS (which stands for Communications, Arts, Technologies and Societies), and the already-mentioned PEACOCKS (which stands for Passion, Ethics, Application, Creativity, Openness, Curiosity, Knowledge and Sustainability).

The merging of the faculties of humanities and social sciences into one also brought a very pleasant surprise. As we predictably had to move into one building, we just happened to occupy the most beautiful one available: the oldest academic premise ever built in Kaunas in 1927. Fully renovated, the new Faculty was now located in A. Mickevičiaus street, 37. A large room on the first floor, no. 111, was assigned to the institute. We now had a beautiful place where to lay our hats. Us being us, we could not help adding a touch of creativity: we hanged congress posters and artworks on the wall, we added an eccentric touch to our noticeboard by framing it with a baroque golden frame, and most of all our junior researcher Ulrika Varankaitė (whom I shall soon mention again) covered our furniture with a lovely decoupage work of semiotic books' pages. We brought in a few accessories, including some board games and a moka machine (which—at least so we believe—makes the best coffee in the whole faculty), and we still occasionally add a thing or two, whenever we have the chance.

Our cherry on top, however, was found during the renovation works of the building. Audra and myself had visited the premises to get an idea of the space and measures: while we were there, we took a walk through the whole building. On the second floor, in a small room, here it was: an old, agonizing and totally-out-of-tune vertical piano in desperate need of adoption. "Can we get that?", I soon asked Audra. A short call to the infrastructure manager gave us the answer we wanted: nobody was going to use that decrepit piano anyway, so we could have it there and then! Once brought into room 111, the second question arose: can we save it? Can it be a piano again, or will it just be a cute piece of furniture (as pianos always are anyway). By all means, this was a job for Motiejus Bazaras, a local young musical genius who, besides being a great pianist, has also a piano assistance and repairing firm. Motiejus came with his toolbox, took a short look and gave us the diagnosis with the same tone by which a competent doctor would say "you've got cancer but we caught it in time". The piano was in a terrible condition: a Soviet instrument that was never meant to compete with a Bösendorfer anyway, it would have still done a fair job if it was not exposed to humidity, little playing, rodents (I did not know they enjoy eating piano parts, actually), and a few other agents that had made it the equivalent of Guevara's and Granado's *La Poderosa* motorcycle after a few kilometres of their legendary journey.

But. But—Motiejus said—the piano can be saved and restored. He literally dismantled the whole thing, and took every single piece in his workshop, leaving only the skeleton in the institute. After a few weeks, the piano was like new, and quickly became a defining feature of room 111. A recent addition was an acoustic guitar, which I found in a second-hand shop: again, not exactly a Martin D-28, but it does its job. We can make music, we have board games and we make good coffee: we have all that is needed to be a research institute, I am sure you will agree.

4.6 Building a Team

The news we received only a few weeks after the official opening of the institute was going to be crucial. We had just been granted an important 3-year research project on "Music and Politics" by the Research Council of Lithuania. Brilliant! My deal with KTU was that we had to earn more staff units through projects. Now we had just done that: we had won a project and we had more staff units. There was now enough to start an actual team.

Both Audra and I were fully convinced that we needed a good combination of professional skills and human qualities: we wanted good people who were also nice people. And, as a bonus, *eclectic* people whenever possible. We figured we could hire three part-time researchers at this point.

An obvious choice would have been my wife Lina Navickaitė-Martinelli, who—family bond aside—is one of the best colleagues I know, and a very eclectic one, for her ability to do well in research, administration, planning, editing/proof-reading (something I am *very* bad at, and I mean "very") and a lot more. However, alas, the Lithuanian laws are very strict on questions of "conflict of interests": members of the same family cannot work together in this kind of way. I of course understand where this rule comes from, and why, and of course I am fully aware (as I already mentioned earlier in this Appendix) of the many frauds of this kind occurring within the academic community. No one wants the immoral *cliché* of a husband hiring his wife (or vice versa) to ensure her a position, no matter the quality of her work. I had after all graduated in a study program, Musicology in Bologna University, where there had been (and still are) a few cases of this sort, much to the detriment of really outstanding scholars, who never managed to achieve the position they deserved, due to this family business. At the same time, a thorough reflection on this "conflict" should be activated. The history of science is full of successful couples of teams of family members, who performed brilliant and groundbreaking research in their fields, never suffering for a second for this situation, and in fact benefiting from it (some of such couples are mentioned in this book as well: Allen and Beatrix Gardner, Thomas and Jean Sebeok, etc.). Family members not only share the same inclination to favor each other: they also share the same problems, and that is often forgotten. By sharing the same problems they can develop more effective strategies to solve them. If they have a small kid (as my wife and I do, at the time I am writing these lines), it is easier to attend to his/her needs without compromising a working day for at least one of them. If they need a working meeting, they can use their dinner table or their home sofa at any point of the day. If one of them is behind schedule, it is easier to make a joint effort to catch up. And so forth. I am not advocating to actively ignore all the negative implications of the "family affair", but I still suggest that meritocracy should remain the number 1 factor by which an employment decision is taken. Maybe mediation between these two opposites can be found.

At any rate, the choice fell on three talented researchers we had met in different situations. One, Dr. Rima Povilionienė, was a colleague in musicology I had known for a long time. She had numerous and interdisciplinary interests in the musicological area, including mathematical issues (thus relating to natural sciences), organology (thus relating to technology) and press representation of music (thus relating to social life): all very numanistic! Also, she had strong skills as a project maker and editor, assets we very much needed.

The second researcher chosen, in chronological order, was Mindaugas Gapševičius, a very creative artist in the modern sense of the term, with a particular eye on technologies and natural sciences, and a strong interest in post-humanism. Although Lithuanian like Rima and Audra, Mindaugas, being also employed at Weimar University of Arts, contributed to increasing our international profile. Also, Mindaugas soon proved to be a terrific project maker, and he got quickly nicknamed a "serial winner", having so far contributed to generating several successful projects for the institute.

The third choice was more of an experiment, and despite good intentions on all sides, turned out to be a failed one. We thought it would be interesting to hire a junior researcher who would not be from the academic world, but from the IT environment. Paulius Paškevičius, a young talented specialist who had a Start-Up named "Solid Education" in Kaunas, was hired with the goal of helping us develop a significant profile in digital humanities (digitalization of our archives, mobile apps of academic interest, etc.). Unfortunately, and I of course take full responsibility for this, we were not able to sufficiently motivate Paulius, so after a few months we had to conclude our working relationship.

In the meanwhile, my teaching activity, for the department of Audiovisual Arts and the department of Philosophy and Psychology, was continuing as always. My main areas of teaching were musical semiotics, audiovisual aesthetics, film theory and music and media. Basically, that area comprised between music and cinema, including the hybrids in the middle. In every course, I could sometimes notice good, or very good, students. When I say "very good" I do not necessarily mean it in the academic sense: some were very good as artists, some in applied activities related to them… but some, indeed, were very good in the academic sense. Having received so much, in terms of "inspiration and perspiration" (to paraphrase Thomas Edison) from my mentors, I was absolutely determined to give back to young students at least as much as I had received when I was one of them. Of all these young scholars, one, Ulrika Varankaitė, had managed to make the upper threshold from "very good" to "outstanding". She was simply the kind of student that makes a teacher's work very easy: she had skills, working ethics and autonomy, all at once. While working with her as a supervisor on her master thesis (on multimodality and psychology in music), it became clear that the junior researcher vacancy left by Paulius had found a very promising heir. Ulrika became an official part of the staff in January 2015. Only a few months later, she also won the competition for doctoral position in art research, here at KTU, so I am currently pleased to supervise her PhD too.

Good students were not the only people I was (and am) taking note of. During my trips around the world to give lectures or presentations, I would always pay attention to particularly talented and active young researchers, especially those who would show genuine interest in what we were trying to do at the institute. After a few years of academic teaching, I have to say, it has become quite easy to distinguish "genuinely interested" and valid people from those who just try to impress you. Being from Italy, that is, the country with the oldest academic staff in the whole of Europe, I know very well how it feels to be a young researcher full of enthusiasm and skills, and simply not having a single chance to display them, because the market is saturated by what in Italy we call "barons" of academia: senior (or very senior) colleagues with lot of institutional power and no interest whatsoever in paving the way for the younger generation. I had seen too many colleagues of my own generation, of enormous value, disappear from the academic environment because all the doors were closed and no one in the older generation (including their own supervisors) was moving a finger to open them. And I knew very well how important it would be for these young scholars to just be given a chance, even a small one, to display their qualities. I knew they would not only deliver, but also give that extra of enthusiasm and motivation.

Little by little, while meeting these interesting young colleagues, I started filling a folder on my computer with their CVs and articles' pdf's, with the goal of involving them in projects, or even trying to hire them, as soon as I had more staff units available. That folder soon became quite precious, becoming a sort of "best of" archive of the new generation of humanists. In the last Imatra school, as mentioned, I had already met the first two entries of the folder, in the persons of Majid Behboudi and Alin Olteanu. By the end of 2014 I had quite an impressive list, and I soon had the opportunity to show it off, as Jurgita Šiugždinienė, our dean, came to room 111 in need of advice. Would I have any idea of some very good international scholar who would accept being employed part-time in KTU to do some teaching, publication, and help out with the preparation of a new study program in communication studies? Internationalizing their profile has been always an important priority for KTU: needless to say, if you are not a very famous institution, or an institution from a big ("highly performing", as EU likes to call them) country, it is more difficult to be an "attractive" working place for foreign scholars.

Anyway, the time to use my precious folder had come. I spoke to Jurgita and I told her quite frankly that I would have been very happy to provide names of very good and realistic candidates, as long as the agreement was that, once hired, this new scholar would work for the institute.

Deal?

Deal.

Great: about ten minutes later (just the time needed for some printing), Jurgita received a good couple of kilos of printed CVs, and could not believe her eyes. Would all these people *really* come and work for us? Of course: we only needed to ask them. After a thorough selection, in which various faculty members participated, we singled out an excellent profile that had become an entry in my folder during one of my many visits at Torino University. Thanks to my colleague and friend Massimo Leone, one of the most outstanding semioticians of my generation, the relationship between KTU and Torino is excellent. Staff exchange is very active, and so is project cooperation. During one of my lectures there, a very smart and proactive specialist in food semiotics with a high publishing profile, named Simona Stano, had stayed after the seminar to introduce herself. She was clearly one of those to file under "genuinely interested", so I asked her to send me some of her material and a CV (which she obliged on the very same day: another good sign!). I was not in the position to promise anything, except my commitment to do my best at any available opportunity. Once the opportunity came, and she was chosen among the many candidates I had assembled till then, she became the sixth researcher of the institute.

Finally, in early 2016, we learned that our university had placed a competition for five post-doctoral positions. Needless to say, we re-opened "the folder", and advanced the candidature of one of its first entries: that same Alin Olteanu I had met in Imatra in 2013. In the meanwhile, we had had another change in our faculty's leadership (the fifth in less than four years, since I had started working for KTU): Jurgita Šiugždinienė had been promoted vice-rector for studies, but for the first time the new dean was appointed in the name of continuity: the former vice-dean for research, Eglė Butkevičienė, became our new leader, continuing just where Jurgita left, and adding her own (many) ideas. Having already built a very good professional and personal relationship with Eglė, this change turned out to be the least traumatic of the lot. Plus, it impressed all of us at the institute that Eglė arranged a specific informal meeting to get acquainted with all our staff personally. A touch of humanity we all took note of.

Thus, Alin's candidature for the post-doc position was discussed with, and strongly endorsed by, our new dean Eglė: his research was going to be related with education, semiotics, multiculturalism and—specifically—Numanities, something extremely important for the scientific growth of our platform. On March 24, 2016, a short message from the central building informed us that, yes, Dr. Alin Olteanu was one of the five candidates selected for the job. Hooray!

So, by May 2016 (last time I had the chance to review my manuscript), this is the team: Audra Daubarienė, Mindaugas Gapševičius, Dario Martinelli, Alin Olteanu, Rima Povilionienė, Simona Stano and Ulrika Varankaitė. Plus, right before going to print, we received the fantastic news that Aušra Berkmanienė, currently Dean's assistant in our faculty, a person of rare kindness, great intelligence and strong working ethics, will be hired from September 2016 to coordinate the organization of the World Congress of Semiotics (of which, after a short mention above, I will tell more later in this chapter). To Aušra, we cannot but happily grant the status of "special guest star" in our institute.

It is difficult to predict how much we will be able to expand: opportunities to hire more staff aside, there are issues of sustainability. I feel that, in the present conditions, we can aim at a 10-member staff (that is, 8 researchers, 1 administrator and 1 director). If we surpass that amount, we may need a reconfiguration of our structure (more administrators, possibly a larger space, a different set of goals, and —most probably—it will become very hard to avoid getting involved in a study program). Time will tell (Fig. 4.5).

Fig. 4.5 The International Semiotics Institute's staff, as of May 2016. *From left to right* Ulrika Varankaitė, Alin Olteanu, Rima Povilionienė, Dario Martinelli, Audra Daubarienė, Mindaugas Gapševičius, Simona Stano

4.7 Developing the Concept of Numanities

As mentioned before, the first important task of the new ISI was the establishment of a credible heir for the Imatra Summer School. Not an easy task, for various reasons: first, as I have repeatedly stated, the cosiness and welcoming atmosphere of the whole environment there had provided the event with an added human value that was a very difficult act to follow. Eero had taught me the importance of that value, through his kindness and aristocratic manners: it was not an aspect I wanted to give up (among other things, it has always been my opinion that the widespread sadness I had witnessed in the last Imatra congress was chiefly due to the fear of losing this cosiness, more than the, anyway high, scientific value of the event).

Second, there was a question of managing three types of community of participants and their interest in the congress: we had to create something that represented continuity with Imatra, in order to maintain part of the old community's interested (it was obvious we were going to lose some of them anyway, but we hoped to gain the trust of at least some of them); we had to create something new to attract a new community (from other academic environments, and other academic "circles"[1]); and we had to create something attractive for the local scholars in Lithuania. The goal was to attract about 20–30 representatives per community, in order to host a congress of at least 60 participants (the last bittersweet Imatra had hosted around 50 participants, so we figured 60 was a good figure for resuming activities).

Third, the trickiest problem. As already mentioned, the institute was now a humanistic unit operating within a technological university: regardless of the faith I could personally have in semiotics, as a methodological tool to address this complex relationship, I knew that my colleagues and supervisors would not have been fully satisfied with that label only. Also, no need to deny it, the appeal of semiotics as a field (and as a term itself) had lost several points in terms of fashionableness and reputation, in comparison with the golden age of the 1960s and 1970s. I was not in a position to defend at all costs the "purity" of semiotics. Some change was needed.[2]

In fact—it soon became clear—it was not only the congress that needed some rebranding: we probably had to think of a whole new research concept, some kind of umbrella that could contain at the same time an International *Semiotics* Institute, the main missions we had set out to pursue (and which I mentioned in the paragraph

[1]I define an "academic environment" as a community interested in a certain area of inquiry, for instance semiotics. By "academic circle", instead, I mean a community whose common denominator is not necessarily a discipline but a set of "contexts": scholars who attend the same events, publish in the same journals, establish the same networks, etc.

[2]To be honest, I was also repeatedly asked to give an introductory seminar about semiotics to my colleagues, and that—in the land of Greimas—may sound a little dodgy. A while later, however, during a brainstorming session aimed at the rebranding of the new faculty, the amount and type of inputs that my semiotic background allowed me to offer to my colleagues was such that I was able to proudly comment: "And this, my friends, is what a semiotician can do". No one asked about that introductory seminar anymore.

"The new era officially begins"), the different research orientations (and inclinations) of our new staff, the demands from KTU in general and the Faculty in particular.

And now, dear readers, a little flash-forward, from the first International Congress of Numanities ICoN (Kaunas, on 2–7 June 2014), on which we shall of course return later:

"Numanities… New this, new that, new, new, new… why does everything always need to be new? Don't we have enough interesting things in this world already?". In Italy, we have a common saying that goes "si dice il peccato, ma non il peccatore": you may reveal the sin, but not the sinner. Therefore, I shall not disclose, here, the name of the very well-known colleague who approached me with these words, on the occasion of the first ICoN. That case proved to be not an isolated one, and since then, in a midst of generally positive feedback and specific compliments for the portmanteau as such, it has indeed been remarked to me that this obsession for novelty at all costs may do more damage than good to the humanities. It was, and is, a legitimate criticism, because, indeed, if we argue so vividly on the necessity of developing the humanities as "appropriate technology", then we should not create a formulation that promotes only "novelty" as a model of progress.

What is altogether amusing, though, is that the very creation of this portmanteau was not exactly animated by the wish to represent "novelty", but rather "conceptuality". I had two sources of inspiration in mind, when the idea popped up, and of course—given my academic and personal background—they had to be both musical. The first source was an excellent electronica band from Finland, Nu Science: I not only liked their music, but I also had the pleasure to be friends with one member, the keyboard player and producer Mikko Ojanen. Nu Science were also the subject of one essay I made on error aesthetics, and they are also briefly mentioned in the first case study of this book. What was charming about this band, and relative name, is that they were not radical innovators (as one would expect from such a name), but in fact their artistic project was mostly *retro*. Enough time has passed since the birth of electronica in popular music: discussing, say, the contribution of Kraftwerk to the development of this genre, means to discuss something that had already happened almost 50 years ago. Nu Science employed analogue instrumentation, including sound cards from early computers, such as the legendary Commodore 64; plus, they composed music that could have been easily suitable for a 1980s videogame, or would have cut a very dignified figure in any of the albums released by German masters like Krafwerk themselves or Tangerine Dream in the 1970s.

In other words: Nu Science were not "nu" (new) in a conventional sense: they were representing an old idea of "new", an idea that had the sound of innovation in a context that by now is not innovative anymore. And at the same time, their compositions are new, they are not covers (although they did nice covers as well: I remember a whole Oasis project, with vocoder-based versions of "Supersonic", "Champagne Supernova" and others), and they are artistically aware of what has been happening in electronic music in the most recent times. Their sound was "retro", not "outdated". In other words, I was enjoying the music of a band that would not use the word "nu" for "novelty in principle", but which was more interested in addressing a historicized dialogue between "old" and "new". I found that very appealing and relevant for the concept that later became "Numanities".

The second source is definitely better-known, although not astoundingly so, if we consider that it was an idea by John Lennon and Yoko Ono, and yet not the most famous of their various initiatives. After a peak of political activism between 1969 and 1972, the couple was little by little focusing on their private life and problems (which were not few, in that period, including a marriage crisis that developed into a separation of one year and a half, ironically nicknamed "The lost weekend" by Lennon himself), and so was their music. On April 1, 1973 (April fool's day), the two launched their last political/pacifist campaign, which—like the previous ones (bed-in, bagism…)—was based on unconventional artistic marketing techniques (most of which was inspired by their most recurrent and not-too-often acknowledged source: the artistic group *Fluxus*). The name of this new campaign was *Nutopia*, as in New Utopia. Nutopia (the Lennons announced in a press conference in New York on that very day) was a "conceptual country", with no boundaries, no laws ("except than cosmic"), no passports—just people. One could become citizen of that country by simply declaring their awareness of it. The country's flag was entirely white (as a metaphor of "surrendering to peace", rather than fighting for it), and its seal was—well—a seal (the mammal). Lennon also "wrote" the Nutopian anthem (releasing it on his 1973 album *Mind Games*): it consisted of 5 s of silence. History has it, this idea was not as successful as the previous ones Lennon and Ono had launched (especially the bed-in, and the related song "Give Peace a Chance", both soon to become absolute icons of pacifism). Yoko Ono still mentions "Nutopia" in some of her pacifist statements, but all in all the idea has remained mostly within fans' knowledge.

However, there were several traits in that idea that attracted me, and which I could relate to in the creation of Numanities. First of all, the fact that the idea was highly conceptual, and well thought of in many aspects (flag, seal, citizenship, etc.). As I mentioned above, the establishment of Numanities was a response to the need for a research "concept", at the International Semiotics Institute. We were not looking for a new discipline (I hope I have repeated it enough times by now), but rather we wanted a single big strategy to cope with different problems at the same time: we were importing from Finland a research institute focused on an area (semiotics) that Kaunas Technological University had never experienced before; we had to be a humanistic unit within a technological institution; we wanted to have a common denominator for our research staff (which was coming from different areas of inquiry). And, sure enough, we not only wanted to have a concept, but we wanted to be conceptual too: we were, after all, rebranding the institute.

Also, like with Nu Science, the idea of "new" conveyed by Lennon's initiative was not of the "novelty at all costs" type. It was in fact one of the oldest ideas ever: the abolition of borders and boundaries and the promotion of global brotherhood (not by chance, the reference is to Thomas More's *Utopia*). Even for Lennon himself, the idea was at least two years old—"Imagine there's no countries" being a concept he famously had sung in 1971.

Finally, the idea had of course a strong ethical emphasis, and it was a vision of progress as something oriented towards dignity of life. Perhaps, a bit of a hippie way to dignity of life, but nevertheless one we can all relate to.

One aspect that we also enjoyed in replacing an "H" with an "N", and that was not inspired by any of the two mentioned sources (or any other, for the matter), was that, by doing so, the root "human" got a little concealed. We did not mind that within the framework of the post-humanistic dimension of the project: we could no longer limit our focus to the human being as such, thus "N" could stand for Nature, too.

And to conclude, while we are at post-humanities, we did not want any "post-" formulation in the creation of the word. We did not want something that comes "after" humanities, because we did not think that humanities had expired. We wanted to add something "on" humanities, something that would represent reformation, but also continuity.

So Numanities were born: out of sources of inspiration that had little to do with the "new this, new that" criticism that my famous colleague had addressed. One cannot blame him, though: if we use "new" as the key-word, we cannot really expect otherwise. And to an extent, we do not mind Numanities to be also "new" humanities *tout court*: we certainly want a degree of innovation—in fact a lot of it. It is just important to reiterate that what we really want is a progress that goes "towards", rather than "forwards" or "backwards": we want "appropriate" technologies, rather than "new" ones; we want novelty that represents evolution, and not novelty in the commercial sense (as, for instance, in the endless spiral of upgrades to a smartphone, that persuades people to repeatedly buy new models, without really having made full use of the previous ones). Whenever "new" stands for these values, we can be very proud to be "new".

4.8 The International Congress of Numanities

With this kind of spirit, a decision was taken to call the new editions of the Summer School "International Congress of Numanities" (ICoN: another manifestation of our obsession for acronyms, with the obvious reference to Peirce's "icon" concept).

As the Summer Schools had always taken place in June, this first edition got scheduled from Monday 2 June to Saturday 7 June, 2014. The theme for the first edition had to be particularly significant within the framework of our new goals and tasks: we chose "The role of humanities in contemporary society: semiotics, culture, technologies". Updating technologies and methodologies, rethinking their role in the academic context, recovering interest and scholarly excellence, adapting (or reacting) to numerous social changes… these—we thought—were all questions that humanities and semiotics alike needed to address in order to reinforce their identity and explore new potentials (Fig. 4.6).

Fig. 4.6 Two moments of the opening ceremony of the first International Congress of Numanities, June 2014

With the same spirit of everything else we were doing in the new ISI, the congress, too, presented elements of innovation and continuity alike, as compared to the Finnish experience. Among the former, logistics and "Numanities" aside:

1. The ICoN Awards, two prizes given respectively to senior and young researchers (with a special five of them distributed in 2014, to kick off the tradition: Eero Tarasti and Philip Tagg in the senior category, Miglė Bisigirskytė, Eglė Gelažiūtė and Ulrika Varankaitė in the junior one);
2. A charity initiative in support of a local orphanage (we collected donations, mostly in the form of clothes and toys).
3. The "Book Fair Trade": a free exchange of semiotic books among participants, based on the "gift economy" principle: a book for a book;
4. The "OpLab!" workshops: an unusual form of session, where semiotic and interdisciplinary knowledge is offered in applied fashion in the form of an "Open Laboratory" accessible not only to congress participants. In 2014, we proposed an OpLab! about Commercial Semiotics and Cultural Analysis, conducted by three semiotic consultants active in three different continents: Dr. Klaus Bernsau from Germany, Dr. Ricardo de Castro Monteiro from Brazil and Dr. Seema Khanwalkar from India;
5. A differentiated form of post-congress publication. Due to the academic institutions' increasing requirements for refereed and indexed articles, our idea was

to offer different publication outlets, rather than a single (usually non-refereed) proceedings volume. While the 2014 edition saw a different range of destinations for the participants' contributions (three journals and one edited collection), the great novelty introduced in 2015 (of which I am about to say) perfected this idea in the best of ways;

6. A differentiated system to pay the registration: instead of having a single fee, the amount changed depending on the Cost of Living index, allowing participants fee to correspond to the proportional living standards of their country.
7. The online streaming of the whole opening ceremony and all the plenary lectures of the congress, thanks to KTU IT department.

Community aside (semiotics and semioticians), important elements of continuity were the format of the congress (separate and parallel sessions, two plenary lectures per day, etc.), a rich social program, and the hosting of SemTra, an "event within the event" symposium on semiotics and translation, which had been part of the Imatra schools for nine years, and was therefore celebrating its 10th edition in ICoN (with Pirjo Kukkonen and Ritva Hartama-Heinonen as its directors).

As the call for papers approached its deadline, and the abstracts were reviewed, we figured out how many colleagues would participate in the first ICoN. We were hoping to reach a quota of 60. We were wrong. We got exactly twice as many: 120 active participants, coming from 21 different countries and 61 academic institutions. Our estimation to get a good sample of each of the three communities we were hoping to attract had an equally-distributed boost: around 40 from the old Summer Schools, around 40 from Lithuania and finally around 40 newcomers. Among the most "numanistic" sessions we managed to organize, I should mention at least the ones entitled "Semiotic Aspects of Possible Future and Ethical Choices", "Cognitive Semiotic Approaches to Cultural Phenomena", "Science-Humanities, Nature-Culture: Boundaries or Bridges?", "Arts and Interdisciplinarity" and "Humanities and/in/for Society". As for the plenary lectures, it seemed to me that the ones most closely incorporating the spirit of what we were trying to do were Dr. Seema Khanwalkar's paper called *Humanities in the Digital Age*, Dr. Roberto Marchesini's *Posthumanism and the Role of Non-Human Alterity in the Humanities*, and Dr. Peter Stockinger's *Digital Audiovisual Archives, Culture and Communication*. Not incidentally, as the perspective of opening the present book series for Springer became more and more concrete, I made sure to approach these three colleagues and ask them to become members of the scientific board.

The cosiness factor, as I mentioned, was a priority for us, and we did our best to match the Imatra standard particularly in the social program, which included two receptions, a rich opening ceremony featuring performances from the Folk Song and Dance Ensemble "Nemunas" and the pianists Eila Tarasti and Rokas Zubovas (the latter underlining the ideal transition from Finland to Lithuania by playing Sibelius's and Čiurlionis's music respectively), a movie evening, a concert given by some of the congress participants, and—perhaps cosiest of them all—a jam-session at a very cool cultural association in Kaunas called "Largo", run by a young and enthusiastic couple, Eglė Gelažiūtė and Tadas Pranevičiūs.

At the end of the week, we had the best-attended event in the whole institute's history. At the end of the month, we closed our budget barely in the black (by a couple of euros, really). At the end of the year, KTU gave us a prize for the best event of 2014.

ICoN2014 had proved to be a very hard act to follow.

In popular music it is commonly maintained that the real parameter for assessing the qualities of an emerging band is not their first album, but rather the second one. If, indeed, the first release gains the audience's attention (when it does) for its freshness, spontaneousness, and largely relies on the surprise/novelty factor, the second release, on the contrary, is expected to be a display of maturity, a confirmation that the promising new band is not a so-called "one hit wonder".

We did not know whether the comparison could go any further than this, but it was undeniable that in preparing the second ICoN we felt similar pressure. The do's and don'ts of the 2015 edition had to be inevitably related to our reflections upon the first one: what seemed to work, what seemed to need improvement.

On the road to the latter, it helped that, during the summer of 2014, the often-mentioned change of headquarters had finally occurred. Now, our office, our faculty and our congress venue could all be in one building. And a beautiful one at that.

Something that had not quite worked as we had hoped were the dates: a very suitable period for Finnish institutions, the first week of June was quite a challenge for Lithuanian ones, as it is the most common period for students' exams. Some Lithuanian colleagues, or so we were told, were not able to participate because of that. The 2015 edition was planned for the last week of May.

Having established a pattern with a three-word theme in 2014 ("Semiotics, Culture, Technologies"), in the call for 2015 we suggested "Creativity, Diversity, Development". In our process of analysis of the state of the art of humanities, this time we wanted to focus on those particular processes where the notion of creativity would emerge in its various forms (artistic expression, social necessity, strategy of survival, scientific methodology, etc.), and would particularly allow social and cultural development within contexts of diversity, minority, otherness, outsideness and originality. Creativity generates diversity generates development. And the other way round.

We confirmed most of the innovations introduced in 2014, so we kept our ICoN awards, our charity project, our BookFairTrade, the online streaming of the opening ceremony and the plenary lectures, the cost of living-based enrolment system, and the Op-Lab! workshop. The latter, this time focused on "user experience", was going to be organized by Majid Behboudi, who had done so much during the first steps of the new ISI. To him, for the junior category, and to the symposium SemTra (in the persons of Pirjo Kukkonen and Ritva Hartama-Heinonen) for the senior category, we thought it appropriate to bestow the 2015 edition of the ICoN Awards.

The social program, besides the very obvious confirmation of the evening at "Largo" (which participants had just adored), featured a celebration of Jean Sibelius's 150th anniversary with a special lecture-concert by Eero and Eila Tarasti and their musicians; a night at the Kaunas City theatre with a representation of Max Frisch's *Biographie*, and the opening ceremony with the Bazaras Brothers jazz trio,

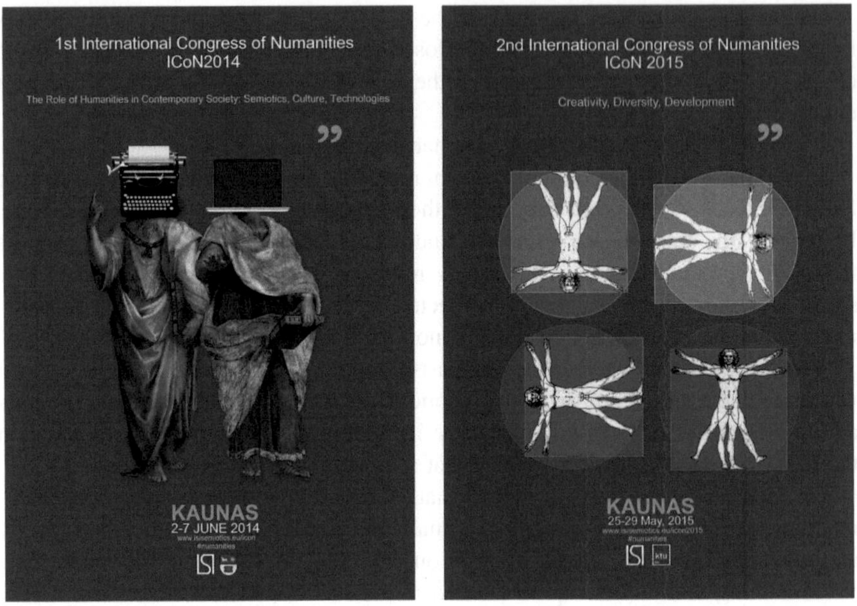

Fig. 4.7 The posters of the 2014 and 2015 editions of the International Congress of Numanities

and two piano performances of Lithuanian and (this time) Polish classical music (the latter, too, meant to become a little tradition: every year we would homage one participant country with their music—Poland was a fair choice to start, as the participation of Polish colleagues in ICoN had been very generous in both 2014 and 2015).

What needed no improvement and yet managed to be so, was the response to the call, which in 2015 reached a record of more than 140 participations, with over 60 academic institutions represented, and almost 30 different world countries. With the previous year's experience in mind, we had the opportunity to arrange even more "numanistic" sessions than before: "Between arts and science", "Disabilities, challenges, potentials", "Politics, social space, relations", "Education, knowledge, growth", "Music, culture, cognition", and others, including two specific workshops on the use of digital technologies for educational purposes organized by the Universidad de las Fuerzas Armadas in Ecuador (Fig. 4.7).

4.9 The Book Series

Before such a big crowd, the occasion was ideal to launch the biggest news for Numanities. Back in 2010, I had published a monograph on zoosemiotics with the Berlin/New York publisher Springer. It had been my most prestigious publication to date, and by far my most pleasant experience with any publisher I had encountered, their staff being a rare combination of professionalism and kindness. I knew that if I ever wanted to embark on a serious editorial project, Springer would have been my first choice. With that spirit, in 2014, I got in touch with Lucy Fleet, responsible for the Philosophy thematic area at Springer, and eventually with the assistant editors Stephen O'Reilly and Silvie Demandt, presenting a project for a whole book series on Numanities. After a few months of communication, and some important adjustment in my proposal (one of the many reasons I should be grateful to Lucy), Springer gave me the great news, right in time to announce it at the congress: they did! They did accept my proposal, and the series would be launched in 2016, under the title "Numanities—Arts and Humanities in Progress".

It was very important, at the stage of assembling the proposal, to be able to count on the support of a sheer all-star team, when it came to forming the scientific committee of the series. In alphabetical order, Luis Emilio Bruni (Aalborg University, Denmark), Paulo Chagas (University of California, Riverside), Teo Forcht Dagi (Harvard Medical School), Kevin Holm-Hudson (University of Kentucky), Gisela Kaplan (University of New England, Australia), Seema Khanwalkar (Ahmedabad University, India), Roberto Marchesini (Center of Study of Posthuman Philosophy, Bologna, Italy), Ricardo de Castro Monteiro (Universidade Anhembi Morumbi, Brazil), Juha Ojala (University of Oulu, Finland), Roger Parent (University of Alberta, Canada) and Peter Stockinger (Institut National des Langues et Civilisations Orientales, Paris, France), all climbed aboard. For the editorial committee, too, I needed a very solid and talented team, and at the same time it was also important to find colleagues I could be more often in touch with (given the number of practicalities needed in their function), so in the end I decided to assemble a group with a relative majority of Lithuanian scholars. The list ended up including Eleonora Adorni (Center of Study of Posthuman Philosophy, Bologna, Italy), Oana Andreica (Music Academy of Cluj, Romania), Audronė Daubarienė (Kaunas University of Technology, Lithuania), Lina Navickaitė-Martinelli (Lithuanian Academy of Music and Theatre, Lithuania), Alin Olteanu (University of Roehampton, UK), Rima Povilionienė (Kaunas University of Technology, Lithuania), Simona Stano (University of Torino, Italy), Natalya Sukhova (Moscow State University), John Tredinnick Rowe (University of Exeter, UK), and Ulrika Varankaitė (Kaunas University of Technology, Lithuania).

Generally speaking, it was agreed with Springer that the series would produce, at least for the beginning, two types of publications: monographs (in-depth studies of a specific topic, with a particular interest in empirical and innovative approaches) and "state-of-the-art" readers/collections (thematic compilations of essays where a given topic or context is approached through a number of recent, innovative studies). What you, dear reader, are now holding in your hand is the first installment of the series.

4.10 Other Projects and Events

Besides ICoN, the institute has organized also events on a smaller scale, mostly meant for the local community: Mandelathon, HowTo Things, and VeGandhi. Their impact on the public, I shall say, has not been awfully encouraging, with the exception of VeGandhi (which has been a moderate, although not smashing as I was hoping, success). I have already mentioned the Mandelathon, which was organized both in 2013 and 2014 in a similar form (a public lecture about Nelson Mandela plus the screening of a documentary and a fiction movie about him). Despite good publicity and encouragement from our dean's office to arrange similar events, both editions were poorly attended—and mostly by those students who were kind of forced to (as assignment within their courses). To motivate students and colleagues to cooperate and participate to the activities organized by any unit was and remains one of the biggest problems that our university has, and one that we constantly address (without significant results, so far) in our faculty meetings.

There are some explanations for this phenomenon, although, perhaps, they are not so convincing to actually become justifications. On the one hand, we are all overloaded with work, and exposed to a considerable amount of stress and time constrictions: the last thing we want is to spend even more of our time with whatever some other unit is organizing. On the other hand (and I shall admit this is totally my "outsider" perspective as a non-Lithuanian), there seems to be a bit of a Soviet cultural-cognitive heritage in that certain indolence not to do anything "extra", because that extra is not paid for. Why doing anything more than the strict necessary, if no one financially (or any other way) rewards us for doing it? The process also becomes a vicious circle as each unit/group gets frustrated by the lack of participation in their event, and actively decides to boycott the next one, whoever is organizing it. I know for a fact that this happens, as I myself did so with the students (not that I am proud of it—but neither do I regret it): after the second melancholically-attended Mandelathon I got so furious that I did not join a big party that the student union had organized in honor of their teachers. I was frankly not in the state for partying with a bunch of young deserters. It turned out I was not the only one to have this idea. During the next faculty meeting, our dean reproached all of us for not attending that party. The head of the student union had been reportedly in tears for having done so much to celebrate us, and having found out that nearly none of the teachers had bothered to go.

I admit it: on my face, I had that sort of devilish smile that Jack Nicholson would be proud of me for. I let the dean and the colleagues know what I thought about this issue: I was sorry about the students, but I was also angry with them, not specifically for the Mandelathon as such, but in general for constantly complaining about the lack of special/significant events in the faculty—and, then, when we actually organize them, showing that they cannot care less. If partying-partying-partying is what they call "special" or "significant" event, then—sorry—a university is not the best place for that.

More or less in the same period as the second Mandelathon (a few days before it, in fact: December 2, 2014), the institute organized what we may call a "quintessentially numanistic" event, the symposium on transdisciplinary practices "HowTo Things". This was the final result of a project that our researcher Mindaugas Gapševičius had won with the Lithuanian Cultural Council. The project departed from the lack of information in relation to the application of knowledge of other disciplines, and therefore was meant to bring together different methodologies, coming from the most diverse fields of inquiry (mostly in social and in natural sciences), with the aim of creating an online instruction guide ("How to…", indeed) for research practices that work with and combine different disciplines. The collected material would then be systematized and presented in easy-to read language, as a set of "tools" that scholars could eventually employ in their own research. The first stage consisted of collecting a number of interviews and profiles from different scholars, who would illustrate their research and explain their methods in a way that could be abstracted from the actual research topic, and become "usable" material. In the second stage, Mindaugas created a Wiki site (the actual online guide), where texts and videos of these researchers were assembled, and that everybody (in a "Wiki" sense) could update and upgrade (www.howto-things.com: it is still there, and it welcomes more contributions). The third and final stage was going to be a symposium, where some of the researchers previously interviewed, plus others, would have their papers.

Although extremely successful in its scientific outcomes, the symposium was again unsuccessful in terms of audience and attendance, and this time—more than students—it was colleagues who had to be blamed. Every single Lithuanian institution related with academic research (from the ministry to the research council, up to each university and academy) is making a big deal of the need to encourage and practice inter-, multi-, cross- and transdisciplinarity. We scholars (I already mentioned it in the first part of the book) enjoy defining ourselves and our research as interdisciplinary (or multidisciplinary, or transdisciplinary… whatever: as I said, by now these significantly different words are trivialized into synonyms), but we rarely do it for real.

Therefore, an event of this sort was meant more for experienced scholars, than for students. Yet, our auditorium was again semi-empty. Rome was not built in a day, better luck next time, et cetera.

The need for more cooperation (and support, and solidarity) within the faculty had become at that point obvious to everybody, but every unit—it seemed—was only ready to claim it, not to offer it. It was therefore an important step forward, a few months later, the one that saw our institute involved with KTU's Municipal Training Center, in the latter's project entitled "Volunteering-Code of Active Citizenship (V-CODE)", funded by EU's Europe for Citizens Program (and counting, a.o., on the important collaboration with Prof. Thomas Bryer, from the University of Central Florida). The aim of the project was to build European partnerships in order to promote, support and encourage active citizenship and volunteering. Being a very important project, the center's director, Jolanta Vaičiūnienė, had asked faculty colleagues to "seriously" cooperate on this project, with initiatives of different sorts: the reaction had again been lukewarm, but this time our dean was determined to make something happen. An idea was developed to make the whole faculty involved in a volunteering promotion: 2015–16 was declared the "VoluntYear". Various initiatives were organized, and the institute participated in four of them: a public lecture, a video assignment for students (who were asked to release documentaries about NGOs in Kaunas), a public film presentation (*The Motorcycle Diaries*, the story of the young Ernesto Guevara volunteering as a young doctor in South America), and then—we all agreed—it would be nice to make a little symposium on the type of Mandelathon, maybe a third edition of Mandelathon itself, but with more focus on the "volunteering" aspect (anti-racist activism, etc.).

Or—we suggested—we could think of another important historical character, Mahatma Gandhi, and thus target another important form of active citizenship that "VoluntYear" had not touched, so far: animal rights, and in particular veganism—which also addresses the question of using eating (or not eating) as a basic way of affecting a society and its dynamics, both at a pragmatic and symbolic level. Veganism is gaining increasing attention in the public debate, yet, the topic has not yet trespassed the gates of academic discussion, the same way as—say—feminism (through gender studies), pacifism (through peace and conflict studies) and others have. Laura Wright has coined the expression "Vegan studies" to define an area that examines the social and cultural discourses on veganism as an identity category and/or social practice.

In addition, we discussed with Jolanta and her team, as more and more public personalities were endorsing veganism: there was a good chance to get some Lithuanian VIPs interested in participating and sharing their own experience on the topic. This could give an unusual (and certainly very attractive) flavor to the event, mixing academic elements with more popularizing ones. And, of course, we could invite some volunteers from animal rights associations and listen to their experience as well.

VeGandhi developed from that conversation. It was going to be the very first academic event on vegan studies in Lithuanian history, and certainly one of the first in absolute terms. The public personalities responded enthusiastically, particularly from the music business, and in the end four of them participated (with no request for honorary, or things like that, it must be said): Jurgis Didžiulis (singer-songwriter), Edgaras Lubys (better known by his stage-name Amberlife, singer-songwriter), Erica Jennings (singer-songwriter), and Jurga Lago (jewellery artist). In addition, we also invited the volunteers Brigita Kymantaitė (from 'Animal Rights Protection Lithuania'), and Gabrielė Vaitkevičiūtė (from the association 'Empty Cages').

Everything went smoothly and efficiently in this case, and finally the public's response (probably—no need to deny it—thanks to the attraction exercised by the VIPs) was decent, with about 70–80 participants, plus several hundreds following the online streaming (still available at https://www.youtube.com/watch?v=QzUXHLhrL_Q, for those who are interested). Again, it was not a roaring success either. Our auditorium can contain up to 200 people, so empty seats were still more than visible (and I regret not having followed Jolanta's advice to actually arrange the event in a smaller—but less nice—hall of the faculty, using the classic trick of concert organizers, who move to a smaller venue when they notice that the ticket sales are not as expected, and therefore still manage to give a "sold-out" impression).

Nevertheless, VeGandhi was a really memorable event (Jurgis Didžiulis' exhilarating intervention being for instance a show within the show): it was meant to be a highly informative occasion, with no pretention to "preach" or "patronize" anyone (as sometimes vegans tend to do with people who have not made the same choice in life), but rather to set the record straight in a society, the very meat-oriented Lithuanian one, which tends to spread plenty of misleading information and negative myths about veganism, via the media and some organizations, in particular the ever present macho stereotype that "real men eat meat, vegans are pussies" and the general perception (much beloved by film directors, apparently) that vegans are all weirdos. For all these reasons, we did not want a "militant" event, where people push the point too hard, but rather we wanted to convey a message of the "normality" of veganism. We wanted the audience to understand that people of all sorts, who are not necessarily radicals/hippies, and who in fact can also be successful role models for youngsters, make this choice as a natural consequence of their sensibility, view of society and ideas.

Through VeGandhi, finally, the institute also had the opportunity to officially announce their support for the "Meat Free Mondays" campaign (http://www.meatfreemondays.com/supporter_categories/universities/), becoming the seventh academic institution in the world to do so, and the very first in Lithuania (in absolute, not only within the academic world) (Fig. 4.8).

As I am writing these lines (end of 2015), the institute is preparing the third edition of ICoN, which will be organized from May 23rd to 26th, 2016. The congress theme this time will be "Processes, Maps, Narratives", an idea that stems from the awareness that plenty of humanistic practices seem to have switched their main focus from "result" to "process". This is for instance the case with the arts (where the concepts of "preparation", "making of", "work in progress" are becoming more and more central in the development and implementation of artwork), but in fact it is the whole of society that seems to have adopted an "open kitchen" principle as the predominant way of understanding and experiencing its many practices.

Processes can of course be displayed in space and/or time, and "maps" and "narratives" are the main forms of articulation of the two dimensions. It is apparent that the "mapping" (that is, the visualization, listing, geographicalization) of knowledge has acquired an importance within humanities that was previously evident only in natural sciences. But at the same time, the end of postmodernism has also invalidated the "death of narratives" claim, and storytelling, time unit-based, discursive forms of knowledge that have regained attention that was temporarily neglected in the previous couple of decades. What is thus the current status (and relationship) of maps and narratives? Are they conflicting or

Fig. 4.8 The posters of the ISI "numanistic" events Mandelathon and VeGandhi

collaborative strategies? What are the pros and cons brought by these two "processes", when it comes to the shaping and development of knowledge?

We expect contributions from areas such as mapping studies and narratological studies, of course, but also insights on artistic "processes" (e.g., creative strategies, music production, film direction, "open kitchen" principles, collective art, etc.); methodologies of knowledge development, with a particular emphasis on the time/space construction (as applied to various fields of humanities, semiotics, education, philosophy…); studies on maps and narratives in new technologies (e.g., the employment of the two processes in the creation of a mobile app, the progressive spatialization and visualization of virtual communication…); investigations on the "uncovering"/"opening" of social, ethical, political processes; plus our "usual" interdisciplinary dialogues between humanities and natural sciences, this time within the context of the three key-words (e.g., how mapping approaches in natural sciences have affected humanities, and how narrative approaches in humanities have affected natural sciences).

Last but not least, as I have already mentioned, the institute will organized the 13th IASS-AIS World Congress of Semiotics in 2017. Hosted every two-three years in different venues, the congress will visit Lithuania for the first time, bringing to the city of Kaunas an approximate amount of 500–1000 semioticians from all over the world. With such figures, the congress qualifies as one of the biggest academic events ever organized in Lithuania, and certainly the biggest ever hosted by KTU. At the time I am writing these lines, only two things are clear: the

Fig. 4.9 The logo of the 13th IASS-AIS World Congress of Semiotics, to be hosted in 2017 by the International Semiotics Institute

CROSS-INTER-MULTI-TRANS-

congress theme will be "Cross-Inter-Multi-Trans" (a celebration of "relations", really, which can take the form of dozens of topics and theoretical approaches: from *inter*disciplinarity to *multi*modality, from *trans*lation to *cross*culturalism...), and the logo will be a platypus (see Fig. 4.9): a "cross-inter-multi-trans" animal by definition (*Monotremata* are mammals with bird characteristics), this mascot is of course meant also as a homage to the recently-departed Umberto Eco, author of the book *Kant and the Platypus*.

4.11 The Inevitable SWOT

To conclude this chapter (and in fact the whole book), I could not resist the temptation to do what in general I hate doing when someone else asks me to: the infamous SWOT analysis—the unavoidable accessory of every single academic project, plan or strategy of nowadays. If—as we said at the beginning—we seriously intend to "re-establish the dialogue between humanities and institutions", and if the writing of an "institutional history" of a given project is worthwhile in this sense, then an assessment of the related institution's current organizational strengths (S) and weaknesses (W) and the expected opportunities (O) and threats (T) from the external environment, is a useful way to wrap up the whole discussion. Plus, even if nobody admits it, a SWOT scheme is nothing other than a classic semiotic Greimasian square, easily obtainable by logically combining the A and non-A factors of "Positive" and "Negative" with the B and non-B factors of "Status" and "Possibility". Therefore, a "positive status" is a strength, a "negative status" is a weakness, a "positive possibility" is an opportunity and a "negative possibility" is a threat.

In relation to its general condition and to its connection with Numanities, the International Semiotics Institute, since its new establishment in Lithuania at the beginning of 2014, and counting until the end of 2015 (when I am writing these lines), can be summarized in the following manner:

Strengths

S1. **Reputation/history of ISI**
The institute has a long history (more than 25 years, since 1988), and is already very well known within the community of semioticians.

S2. **Creativity of ISI**
It seems to be the strongest value within the staff, and it is employed in many ways: research, planning, marketing, etc. Moreover, 4/6 of the staff members are specialized in art and/or art research.

S3. **Internationality of ISI**
This is an asset that is particularly significant within KTU and Lithuania (which is growing in this sense, but still struggling with the counterforces of locality and isolation). ISI is the only academic research institute in Lithuania to have a non-Lithuanian director. In addition, four staff members are also affiliated to foreign institutions, three are not Lithuanian, the official language for every initiative/activity of the institute is English, etc.

S4. **ISI is strongly inter-/cross-/trans-/multidisciplinary**
Most of the events organized (ICoN, HOWTO, VeGandhi…) and research produced aim at a dialogue across different disciplines, in various forms.

S5. **Productivity of ISI**
So far, the institute has been able to maintain a rather high profile in the areas of publishing (with a total of over 20 publications in two years, among them articles, monographs and edited collections), project applications (with an average of almost one per month, and a success rate of 20 %), and events (two international congresses and three symposia in two years).

Weaknesses

W1. **Human resources**
At present, ISI can count on only six staff members. Among other problems, this limits the institute's ambition, and sustainability of "think big" strategies.

W2. **Bureaucratic constraints**
KTU is a highly-bureaucratic institution, and that slows down and demotivates several ISI activities.

W3. **Lack of related PhD program**
Humanities, at KTU, have only one doctoral program in Art Research, with only one place per year. None exists in more semiotics-related areas. This impairs the natural development and refreshment of the researchers' community.

W4. **ISI is missing some specific competences**
Due to the small number of staff members, ISI has some important gaps in areas that would be really central in Numanities.

W5. **ISI's limited facilities**
Not being a very big institute, ISI does not possess the biggest choice of facilities, infrastructure and equipment. That makes it more challenging to "think big" in terms of projects and events.

Opportunities

O1. **ISI's innovative research concept and related support from KTU**
Numanities, the subject of this book, is a research concept in growth, whose future cannot yet be predicted in detail, but which seems to be gaining an increasing consensus. In that, it must be acknowledged that KTU was soon supportive to the idea.

O2. **Technology + Humanities**
The peculiar condition of a humanist institute within a technological university seems to have strong potential in the further development of areas like digital humanities, creativity, e-inclusion, etc.

O3. **ISI is one of very few semiotics institutes**
Semiotic organizations exist mostly at the level of associations or study programs. ISI is one of the very few "research institutes" within a university.

O4. **Developing in Greimas' country**
The territory (home to one of the greatest semioticians, Algirdas J. Greimas) may strengthen ISI's institutional position, and let ISI strengthen semiotics' position in Lithuania—creating something similar to countries like Estonia, France or Italy, which—thanks to the respective "national heros" of semiotics (Lotman, Barthes, Eco…)—where semiotics is a very "institutionalized" field.

O5. **Placing Kaunas on the international academic map**
Given the previous points, Kaunas may acquire more international visibility, as an active centre for semiotics and indeed Numanities.

Threats

T1. **ISI's uncoordinated research strategy**
ISI's research profile and related activities are so far based on the individual inclinations and interests of the staff members. While this is generally regarded as a value (the research strategy is bottom-up), if not given a proper framework, this situation may develop into a dispersion of resources and limitation of impact.

T2. **KTU's internationality not in step**
ISI is in general more internationalized (language, events, attitude…) than KTU. That tends to slow down, impair some activities, or at least result in a limited understanding of their potential. Having said that, it must be acknowledged that KTU is rapidly growing in this respect, and it is reasonable to predict that in few years this threat will become an opportunity.

T3. **KTU's business-oriented strategy**
KTU, like other universities, has now an enterprise-like and profit-oriented mentality, which makes it harder for ISI to pursue research/scientific goals, and has also additional bureaucratic constraints.

T4. **KTU's investment**
As a consequence, KTU seems to prefer short-term actions, rather than long-term investments that would allow a unit's organic growth.

T5. **General perception of semiotics**
Though the situation is now again improving, semiotics has still not recovered the popularity level of the 1960s–1970s. That may still damage some activities like project applications (it is often better *not* to mention semiotics in them, for fear of rejection from skeptical reviewers).

References

Allport, Gordon W. 1954. *The Nature of Prejudice*. Reading, MA: Addison-Wesley.

André, Michael, and Cees Kamminga. 2000. Rhythmic dimension in the echolocation click trains of sperm whales. *Journal of the Marine Biological Association of the United Kingdom* 82: 163–169.

André-Alt, Peter. 2012. Humanities in modern society. Speech delivered at New York, German House, April 4, 2012.

Arnold, Kate, and Klaus Zuberbühler. 2006. Language evolution: Semantic combinations in primate calls. *Nature* 441: 303.

Badmington, Neil. ed. 2000. *Posthumanism*. New York: Palgrave.

Bankov, Kristian. 2004. Infinite semiosis and resistance. In *From Nature to Psyche*, ed. Eero Tarasti, 175–181. Helsinki: International Semiotics Institute.

Barbieri, Marcello. 2008. Biosemiotics: A new understanding of life. *Naturwissenschaften* 95(7): 577–599.

Barthes, Roland. 1957. *Mythologies*. Paris: Editions de Seuil.

Bateson, Gregory. 1969. Metalogue: What is an instinct? In *Approaches to Animal Communication*, eds. Thomas A. Sebeok and Alexandra Ramsay, 11–30. The Hague: Mouton.

Bateson, Gregory. 1972. *Steps to an Ecology of Mind*. Chicago, IL: University of Chicago Press.

Battaglia, Luisella. 1997. *Etica e diritti degli animali*. Bari-Roma: Laterza.

Bauer, A. Michael, and Aaron Brown. 2014. Quantitative assessment of appropriate technology. *Procedia Engineering* 78: 345–358.

Bekoff, Marc. 1995. Cognitive ethology and the explanation of nonhuman animal behaviour. In *Comparative Approaches to Cognitive Science*, eds. Herbert L. Roitblat and Jean-Arcady Meyer, 119–150. Cambridge, MA: MIT Press.

Bekoff, Marc, and Dale Jamieson. eds. 1990. *Interpretation and Explanation in the Study of Animal Behaviour*. Boulder, CO: Westview Press.

Bopry, Jeanette. 2002. Semiotics, epistemology, and inquiry. *Teaching and Learning* 17(1): 5–18.

Bouissac, Paul. ed. 1998. *Encyclopedia of Semiotics*. New York: Oxford University Press.

Broadhead, Richard H., et al. eds. 2013. *The Heart of the Matter—The Humanities and Social Sciences for a Vibrant, Competitive, and Secure Nation*. Cambridge, MA: American Academy of Arts & Sciences.

Brosnan, Mark J. 1998. *Technophobia—The Psychological Impact of Information Technology*. London: Routledge.

Brown, Rupert. 1989. *Group Processes: Dynamics Within and Between Groups*. Oxford: Basil Blackwell.

Bruner, Jerome. 1986. *Actual Minds, Possible Worlds*. Cambridge: Harvard University Press.

D. Martinelli, *Arts and Humanities in Progress*, Numanities - Arts and Humanities in Progress 1, DOI 10.1007/978-3-319-45553-2

Byrne, Richard W., and Andrew Whiten. 1991. Computation and mindreading in primate tactical deception. In *Natural Theories of Mind*, ed. Andrew Whiten, 127–141. Oxford: Basil Blackwell.
Cage, John. 1961. *Silence: Lectures and Writings*. Middleton: Wesleyan University Press.
Carlyle, Thomas. 1858. *The Collected Works of Thomas Carlyle*. London: Chapman and Hall.
Cavalieri, Paola. 1999. *La questione animale*. Torino: Bollati Boringhieri.
Cimatti, Felice. 1998. *Mente e linguaggio negli animali*. Roma: Carrocci.
Cimatti, Felice. 2000. *Nel segno del cerchio*. Manifestolibri: L'ontologia semiotica di Giorgio Prodi. Roma.
Clayton, Nicola S., Timothy J. Bussey, and Anthony Dickinson. 2003. Can animals recall the past and plan for the future? *Nature Reviews Neuroscience* 4(8): 685–691.
Correia, Sergio P., Anthony Dickinson, and Nicola S. Clayton. 2007. Western scrub-jays anticipate future need states independently of their current motivational state. *Current Biology* 17: 856–861.
Dahlhaus, Carl, and Hans H. Eggebrecht. 1985. *Was ist Musik?* Wilhelmshaven: Heinrichshofen's Verlag.
D'Andrade, Roy. 1995. Moral Models in Anthropology. *Current Anthropology* 36(3): 399–407.
Darwin, Charles. 1871. *The Descent of Man and Selection in Relation to Sex*. London: Murray.
Darwin, Charles. 1872. *The Expression of Emotions in Man and Animals*. London: Murray.
Deely, John. 2000. The impact of semiotics on philosophy. Paper presented at Helsinki University, 1st Annual Hommage to Oscar Parland, December 1, 2000.
Deely, John. 2005. *Defining the Semiotic Animal—A Postmodern Definition of Human Being*. Sofia: Tip-Top Press.
Deely, John. 2009. *Basics of Semiotics*, 5th ed. Tartu: Tartu University Press.
Dennett, Daniel C. 1996. *Kinds of minds*. New York: Hypercollins.
Dilthey, Wilhelm. 1991. *Selected Works, Volume I: Introduction to the Human Sciences*. eds. Rudolf A. Makkreel and Frithjof Rodi. Princeton, NJ: Princeton University Press.
Divers, John. 2002. *Possible Worlds*. London: Routledge.
Dodsworth, Simon. 2009. *The Fundamentals of Interior Design*. Worthing: AVA Publishing.
Downs, Anthony. 1992. *Stuck in Traffic: Coping with Peak-Hour Traffic Congestion*. Washington, DC: The Brookings Institution.
Dreifus, Claudia. 1998. She talks to Apes and, according to her, they talk back. *New York Times*. Retrieved November 14, 2015: http://williamcalvin.com/teaching/NYTimes-SSR-4-98.htm.
Eagleton, Terry. 2015. The slow death of the university. *The Chronicle of Higher Education*. http://chronicle.com/article/The-Slow-Death-of-the/228991/. Accessed October 2, 2015.
Eco, Umberto. 1987. Fakes and forgeries. *Versus* 46: 3–29.
Eco, Umberto. 1997. *Kant e l'ornitorinco*. Milano: Bompiani.
Elliott, Jennifer A. 2006. *An Introduction to Sustainable Development*. London: Routledge.
Emery, Nathan J., and Nicola S. Clayton. 2008. Imaginative scrub-jays, causal rooks, and a liberal application of Occam's aftershave. *Behavioral and Brain Sciences* 31: 134–135.
Engels, Friecrich. 1896. The part played by labour in the transition from ape to man. *Die Neue Zeit* XIV/2: 545–554.
Epstein, Mikhail. 2012. *The Transformative Humanities. A Manifesto*. London: Bloomsbury.
Esch, Harald. 1967. The evolution of bee language. *Scientific American* 216(4): 97–104.
Evans, Edward P. 1906. *The Criminal Prosecution and Capital Punishment of Animals*. London: Faber & Faber.
Faulkner, Audrey O., and Maurice L. Albertson. 1986. Tandem use of hard and soft technology: An evolving model for third world village development. *International Journal of Applied Engineering Education* 2(2): 127–137.
Ferrando, Francesca. 2013. Posthumanism, transhumanism, antihumanism, metahumanism, and new materialisms—Differences and Relations. *Existenz* 8/2. http://www.bu.edu/paideia/existenz/volumes/Vol.8-2Ferrando.html. Accessed October 2, 2015.

Fisher, John A. 1990. The myth of anthropomorphism. In *Interpretation and Explanation in the Study of Animal Behaviour*, eds. Marc Bekoff and Dale Jamieson, 96–116. Boulder, CO: Westview Press.
Fouts, Roger S. 1997. *Next of Kin*. New York: William Morrow.
Fox, Nicols. 2002. *Against the Machine. The Hidden Luddite Tradition in Literature, Art, and Individual Lives*. Washington: Island Press.
Freud, Sigmund. 1899. *Die Traumdeutung*. Leipzig: Franz Deuticke.
Freud, Sigmund. 1959. *Beyond the Pleasure Principle*. New York: Norton.
Fukuyama, Francis. 2004. Transhumanism. *Foreign Policy* September/October: 42–43.
Galtung, Johan. 1969. Violence, peace and peace research. *Journal of Peace Research* 6(3): 167–191.
Gardner, Allen R., and Beatrix T. Gardner. 1969. Teaching sign language to a chimpanzee. *Science* 165: 664–672.
Gardner, Allen R., Beatrix T. Gardner, and Thomas E. Van Cantfort. eds. 1989. *Teaching sign language to a chimpanzee*. Albany, NY: State University of New York Press.
Gerard, Genette. 1972. *Figures III*. Paris: Seuil.
Goodall, Jane. 1971. *In the Shadow of Man*. Boston, MA: Houghton Mifflin.
Gora, Joseph, and Andrew Whelan. 2010. Invasion of aca-zombies. *The Australian 3/11/2010*. http://www.theaustralian.com.au/higher-education/opinion/invasion-of-aca-zombies/story-e6frgcko-1225946869706. Accessed June 15, 2015.
Gorman, James. 2012. Animal studies cross campus to lecture hall. *The New York Times*. http://www.nytimes.com/2012/01/03/science/animal-studies-move-from-the-lab-to-the-lecture-hall.html?_r=0. Accessed November 14, 2015.
Grossberg, Lawrence. 1992. *We Gotta Get Out of This Place: Popular Conservation and Postmodern Culture*. New York: Routledge.
Habermas, Jürgen. 1998. *The Inclusion of the Other. Studies in Political Theory*. Cambridge, MA: MIT Press.
Halberstam, Judith, and Ira Livingston. eds. 1995. *Posthuman Bodies*. Bloomington: Indiana University Press.
Harvard Magazine. 2013. Addressing a Decline in Humanities Enrollment. *Harvard Magazine*, June 6, 2013. http://harvardmagazine.com/2013/06/reinvigorating-the-humanities. Accessed March 1, 2015.
Hausdorff, Felix (as Paul Mongré). 1897. *Sant'Ilario. Gedanken aus der Lanschaft Zarathustras*. Leipzig: C.G. Naumann.
Herrick, Paul. 1999. *The Many Worlds of Logic*. Oxford: Oxford University Press.
Hill, Jane H. 1980. Apes and language. In *Speaking of Apes*, eds. Thomas A. Sebeok and Jean Umiker-Sebeok, 331–352. New York: Plenum Press.
Hockett, Charles F. 1960a. The origin of speech. *Scientific American* 203: 89–97.
Hockett, Charles F. 1960b. Logical considerations in the study of animal communication. In *Animal Sounds and Communication*, eds. Wesley E. Lanyon and William N. Tavolga, 392–430. Washington: American Institute of Biological Sciences.
Hoffmeyer, Jesper. 1995. The semiotic body-mind. In *Essays in Honor of Thomas A. Sebeok*, ed. Norma Tasca, 367–383. Porto: Almeida.
Hoffmeyer, Jesper. 1996. *Signs of Meaning in the Universe*. Bloomington: Indiana University Press.
Hoffmeyer, Jesper. 1997. Biosemiotics: Towards a new synthesis in biology? *European Journal for Semiotic Studies* 9(2): 355–376.
Holmes, Thomas B. 1985. *Electronic and Experimental Music*. New York: Charles Scribner's Sons.
Horowitz, Damon. 2011. From Technologist to Philosopher: Why you should quit your technology job and get a Ph.D. in the humanities. *The Chronicle of Higher Education*. http://chronicle.com/article/From-Technologist-to/128231/. Accessed March 1, 2015.
Inglis, Sam. 2002. Markus popp: Music as software. *Sound on Sound*. http://www.soundonsound.com/sos/oct02/articles/oval.asp. Accessed September 5, 2015.
Jakobson, Roman. 1963. *Essais de linguistique générale*. Paris: Editions de Minuit.

Jameson, Fredric. 1991. *Postmodernism or, the Cultural Logic of Late Capitalism*. Durham: Duke University Press.

Jevons, William S. 1865. *The Coal Question; An Inquiry Concerning the Progress of the Nation, and the Probable Exhaustion of Our Coal Mines*, 1 ed. London: Macmillan & Co.

Jones, Steven E. 2006. *Against Technology: From the Luddites to Neo-Luddism*. London: Routledge.

Jump, Paul. 2015. Academics in the minority at more than two-thirds of UK universities. In *Times Higher Education*. Retrieved October 20, 2015: https://www.timeshighereducation.com/news/academics-minority-more-two-thirds-uk-universities.

Katkus, Donatas. 2006. *Muzikos Atlikimas*. Vilnius: Lietuvos Muziku Sajunga.

Kessel, Edward L. 1955. The mating activities of balloon flies. *Systematic Zoology* 4: 96–104.

Kivy, Peter. 1995. *Authenticities: Philosophical Reflections on Musical Performance*. Ithaca: Cornell University Press.

Kohts, Nadia. 1935. *Infant Ape and Human Child Instincts, Emotions, Play, Habits*. Moscow: Scientific Memoirs of the Museum Darwinianum.

Kull, Kalevi. 1998. On semiosis, Umwelt, and semiosphere. *Semiotica* 1203(4): 299–310.

Kull, Kalevi. 1999. Biosemiotics in the twentieth century: A view from biology. *Semiotica* 1271 (4): 385–414.

Lewis, David. 1986. *On the Plurality of Worlds*. Oxford: Basil Blackwell.

Liebermann, Philip. 1968. Primate vocalizations and human linguistic ability. *Journal of the Acoustical Society of America* 44: 1574–1584.

Liebermann, Philip. 1972. *The Speech of Primates*. The Hague: Mouton.

Louie, Kenway, and Matthew A. Wilson. 2001. Temporally structured replay of awake hippocampal ensemble activity during rapid eye movement sleep. *Neuron* 29(1): 145–156.

MacDonald, Heather. 2014. The humanities have forgotten their humanity. *Wall Street Journal*, January 3, 2014. http://www.wsj.com/articles/SB10001424052702304858104579264321265378790. Accessed October 2, 2015.

MacNealy, Mary S. 1999. *Strategies for Empirical Research in Writing*. Boston, MA: Allyn & Bacon.

Mainardi, Danilo. 1975. *L'animale culturale*. Milano: Rizzoli.

Mainardi, Danilo. ed. 1992. *Dizionario di etologia*. Torino: Einaudi.

Malacarne, Giorgio. 1992. Linguaggio. In *Dizionario di etologia*, ed. Danilo Mainardi, 451–457. Torino: Einaudi.

Malacarne, Giorgio. 2005. Il linguaggio simbolico negli uomini e nelle scimmie. In *Il comportamento degli animali*, ed. Enrico Bellone, 243–247. Novara: De Agostini.

Mali. Taylor. 2002. *Totally like whatever, you know? What Learning Leaves*. Newtown, CT: Hanover Press.

Marler, Peter. 1977. The structure of animal communication sounds. In *Recognition of Complex Acoustic Signals*, ed. Theodore H. Bullock, 17–35. Berlin: Springer.

Martinelli, Dario. 2002. *How Musical Is a Whale?* Helsinki: International Semiotics Institute.

Martinelli, Dario. 2006. Language and interspecific communication: a case to re-open. In *Introduction to Biosemiotics*, ed. Marcello Barbieri, 473–518. Berlin: Springer.

Martinelli, Dario. 2007. *Zoosemiotics: Proposals for a Handbook*. Helsinki: International Semiotics Institute.

Martinelli, Dario. 2008. Music, identity and the strange case of authenticity. *Lithuanian Musicology* 9: 122–130.

Martinelli, Dario. 2009a. Notes on the theoretical program of biosemiotics; or what exactly am I doing here? In *Communication: Understanding/Misunderstanding*, ed. Eero Tarasti, 1022–1032. Helsinki: International Semiotics Institute.

Martinelli, Dario. 2009b. *Of Birds, Whales and Other Musicians: Introduction to Zoomusicology*. Scranton: University of Scranton Press.

Martinelli, Dario. 2010a. *A Critical Companion to Zoosemiotics: People, Paths, Ideas*. Berlin: Springer.

Martinelli, Dario. 2010b. *Authenticity, Performance and Other Double-Edged Words—Essays on Popular Music*. Helsinki: International Semiotics Institute.

Martinelli, Dario. 2010c. Estetica dell'errore ed 'autenticità' nella musica elettronica – Riflessioni, e il caso dei finlandesi Nu Science. *Lexia* 5–6: 217–254.

Martinelli, Dario, and Kristian Bankov. 2008. Bankov's Razor Versus Martinelli's Canon. A confrontation around biosemiotics. *Biosemiotics* 1/3: 397–418.

Menton, Jane D. 2013. Decline of the humanities up for debate. *Yale Daily News*, September 17, 2013. http://yaledailynews.com/blog/2013/09/17/women-in-the-humanities/. Accessed April 5, 2015.

Miah, Andy. 2008. Posthumanism in cultural theory. In *Medical Enhancement and Posthumanity*, eds. Bert Gordijn and Ruth Chadwick, 71–94. Berlin: Springer.

Miles, Barry. 1997. *Paul McCartney—Many Years from Now*. London: Secker & Warburg.

Mitani, John C., and Peter Marler. 1989. A phonological analysis of male gibbon singing behavior. *Behaviour* 109(5): 20–45.

Mitchell, Robert W. 1990. A theory of play. In *Interpretation and Explanation in the Study of Animal Behaviour*, ed. Marc Bekoff, and Dale Jamieson, 197–225. Boulder, CO: Westview Press.

Mitchell, Robert W. ed. 2002. *Pretending and Imagination in Animals and Children*. Cambridge: Cambridge University Press.

Moore, Allan. 2002. Authenticity as authentication. *Popular Music* 21(2): 225–236.

Morris, Charles W. 1946. *Signs, Language, and Behavior*. Englewood Cliffs, NJ: Prentice Hall.

Morris, Desmond John. 1963. *The Biology of Art*. London: Methuen.

Morris, Desmond John. 1967. *The Naked Ape*. London: Jonathan Cape.

Morris, Desmond John. 1969. *The Human Zoo*. London: Jonathan Cape.

Mulcahy, Nicholas J., and Josep Call. 2006. Apes save tools for future use. *Science* 312: 1038–1040.

Navickaitė-Martinelli, Lina. 2014. *Piano Performance in a Semiotic Key: Society, Musical Canon and Novel Discourses*. Helsinki: International Semiotics Institute.

Nelson, Gary, and Dilip P. Gaonkar. eds. 1996. *Disciplinarity and Dissent in Cultural Studies*. New York: Routledge.

Norris, Christopher. 1990. *What's Wrong with Postmodernism*. Loughborough: Harvester Wheatsheaf.

Nussbaum, Martha C. 2010. *Not for Profit—Why Democracy Needs the Humanities*. Princeton: Princeton University Press.

Papanek, Victor. 1971. *Design for the Real World: Human Ecology and Social Change*. New York: Pantheon Books.

Patterson, Francine G. 1977. Linguistic capabilities of a young Lowland Gorilla. Paper presented at a Symposium of the American Association for the Advancement of Science: "An account of the visual mode: Man vs. Ape". Denver.

Patterson, Francine G. 1978. The gestures of a gorilla: Language acquisition in another pongid. *Brain and language* 5: 72–97.

Patterson, Francine G. 1981. Ape language. *Science* 211: 87–88.

Patterson, Francine, and Eugene Linden. 1981. *The education of Koko*. New York: Holt, Rineheart & Winston.

Peter, Laurence, and Raymond Hull. 2009. *The Peter Principle: Why Things Always Go Wrong*. New York: Harperbusiness.

Petrilli, Susan. 1998. *Teoria dei segni e del linguaggio*. Bari: Edizioni B.A.Graphis.

Poe, Gina R., et al. 2000. Experience-dependent phase-reversal of hippocampal neuron firing during REM sleep. *Brain Research* 855(1): 176–180.

Ponzio, Augusto, and Susan Petrilli. 2008. Bodies, signs and values in global communication. In *Approaches to Communication. Trends in Global Communication Studies*, ed. Susan Petrilli, 113–141. Madison, WI: Atwood Publishing.

Prodi, Giorgio. 1983. Lingua e biologia. In *Intorno alla linguistica*, ed. Cesare Segre, 172–202. Milano: Feltrinelli.

Putt, Archibald. 2006. *Putt's Law and the Successful Technocrat: How to Win in the Information Age*. New York: Wiley-IEEE Press.

Raby, Caroline R., Dean M. Alexis, Anthony Dickinson, and Nicola S. Clayton. 2007. Planning for the future by western scrub-jays. *Nature* 445: 919–921.

Raphael, Daniel. 2015. *Social Sustainability Handbook for Community-Builders*. Denver: Daniel Raphael Consulting.

Regan, Tom. 1983. *The Case for Animal Rights*. London: Routledge.

Richard, Bernard. 1967. Le determinisme de la construction des barrages chez le castor du Rhone. *La Terre et la Vie* 4: 339–470.

Robinson, John G. 1984. Syntactic structures in the vocalizations of wedge-capped capuchin monkeys, *Cebus olivaceus*. *Behaviour* 90: 46–79.

Rossi-Landi, Ferruccio. 1972. *Semiotica e ideologia*. Milano: Bompiani.

Rothschild, Friedrich S. 1962. Laws of symbolic mediation in the dynamics of self and personality. *Annals of New York Academy of Sciences* 96: 774–784.

Russo, John P. 1998. The humanities in a technological society. *Humanitas* XI/1: 14–41.

Schmidt, Benjamin. 2013. *Sapping Attention*. http://sappingattention.blogspot.com/2013_06_01_archive.html. Accessed October 7, 2015.

Schumacher, Ernst F. 1973. *Small Is Beautiful: Economics as if People Mattered*. Vancouver, BC: Hartley & Marks.

Searle, John R. 1975. The logical status of fictional discourse. *New Literary History* 14: 319–332.

Sebeok, Thomas A. 1963. Communication in animals and men. *Language* 39: 448–466.

Sebeok, Thomas A. 1976. *Contributions to the Doctrine of Signs*. Bloomington, IN: Indiana University Press.

Sebeok, Thomas A. 1981. *The Play of Musement*. Bloomington, IN: Indiana University Press.

Sebeok, Thomas A. 1990. *Essays in zoosemiotics*. Toronto: Toronto Semiotic Circle.

Sebeok, Thomas A. 1991. *A Sign is Just a Sign*. Bloomington: Indiana University Press.

Sebeok, Thomas A. 1994. *Signs: An Introduction to Semiotics*. Toronto: University of Toronto Press.

Sebeok, Thomas A. 1998. *Come comunicano gli animali che non parlano*. Modugno: Edizioni dal Sud.

Sebeok, Thomas A., and Marcel Danesi. 2000. *The forms of meaning*. Berlin: Mouton de Gruyter.

Sebeok, Thomas A., and Jean Umiker-Sebeok. eds. 1980. *Speaking of Apes*. New York: Plenum Press.

Seidel, Asher. 2008. *Inhuman Thoughts: Philosophical Explorations of Posthumanity*. Plymouth: Lexington Books.

Shanahan, Murray P. 2006. A cognitive architecture that combines internal simulation with a global workspace. *Consciousness and Cognition* 15: 433–449.

Shanahan, Murray P., and Bernard Baars. 2005. Applying global workspace theory to the frame problem. *Cognition* 98: 157–176.

Sharov, Aleksei. 2001. Umwelt theory and pragmatism. *Semiotica* 134: 211–228.

Sheets, Diana E. 2013. The crisis in the humanities: Why today's educational and cultural experts can't and won't resolve the failings of the liberal arts. *Huffington Post*, July 15, 2013. http://www.huffingtonpost.com/dr-diana-e-sheets/the-crisis-in-the-humanit_b_3588171.html. Accessed October 9, 2015.

Shuker, Roy. 1998. *Popular Music—The Key Concepts*. New York: Routledge.

Siapas, Athanassios G., and Matthew A. Wilson. 1998. Coordinated interactions between hippocampal ripples and cortical spindles during slow-wave sleep. *Neuron* 21(5): 1123–1128.

Siegel, Lee. 2013. Who ruined the humanities? *Wall Street Journal*, July 12, 2013. http://online.wsj.com/news/articles/SB10001424127887323823004578595803296798048. Accessed October 9, 2015.

Singer, Peter. 1989. All animals are equal. In *Animal Rights and Human Obligations*, eds. Tom Regan and Peter Singer, 148–162. Englewood Cliffs, NJ: Prentice Hall.

Snow, Charles P. 1959. *The Two Cultures and the Scientific Revolution*. New York: Cambridge University Press.

Sokal, Alan. 1996a. Transgressing the boundaries: Toward a transformative hermeneutics of quantum gravity. *Social Text* 46(47): 217–252.

Sokal, Alan. 1996b. A physicist experiment with cultural studies. *Lingua Franca* May/June: 62–64.

Sokal, Alan. 2008. *Beyond the Hoax—Science, Philosophy and Culture*. New York: Oxford University Press.

Smith, Carlyle. 1995. Sleep states and memory processes. *Behavioural Brain Research* 69(1-2): 137–145.

Stefani, Gino. 1985. *Competenza musicale e cultura della pace*. Bologna: CLUEB.

Stefani, Gino. 1989. *Musica con coscienza*. Roma: Paoline.

Stefani, Gino, and Stefania Guerra. eds. 2005. *Dizionario della musica nella globalità dei lingaggi*. Lucca: LIM.

Tahir, Tariq. 2010. The irresistible rise of academic bureaucracy. *The Guardian*, March 30, 2010. http://www.theguardian.com/education/2010/mar/30/academic-bureaucracy-rise-managers-higher-education. Accessed September 13, 2015.

Tajfel, Henri. 1981. *Human Groups and Social Categories*. Cambridge: Cambridge University Press.

Tarasti, Eero. 2000. *Existential Semiotics*. Bloomington: Indiana University Press.

Tarasti, Eero. 2005. Vastarinnan Semiotiikkaa: Oleminen, Muisti, Historia – Merkkien Vastavirta. *Synteesi* 1(2005): 2–29.

Terras, Marisa, et al. 2013. The humanities matter! *Infographic*. http://4humanities.org/wp-content/uploads/2013/07/humanitiesmatter300.pdf. Accessed October 9, 2015.

Thornton, Sarah. 1995. *Club Cultures: Music, Media and Subcultural Capital*. London: Polity Press.

Tindemans, Peter, et al. 2003. *The Future of the Sciences and Humanities*. Amsterdam: Amsterdam University Press.

Trask, R. Larry. 1995. *Language: The Basics*. London: Routledge.

Twenge, Jean M., and W. Keith Campbell. 2009. *The Narcissism Epidemic: Living in the Age of Entitlement*. New York: Simon and Schuster.

van Peer, Willie, et al. 2007. *Muses and Measures—Empirical Research Methods for the Humanities*. Newcastle: Cambridge Scholars Publishing.

von Frisch, Karl. 1967. *The Dance Language and Orientation of Bees*. Cambridge, MA: The Belknap Press of Harvard University Press.

von Glasersfeld, Ernst. 1978. Les chimpanzés et le langage. *La Recherche* 9: 725–732.

Waller, Margaret. 1992. The Melancholy Man and the Lady with the Lyre: The sexual politics of genius in early romantic fiction and painting. In *Correspondances—Studies in Literature, History and the Arts in 19th Century France*, ed. Keith Busby, 223–237. Amsterdam: Rodopi.

Wicklein, Robert C. 1998. Designing for appropriate technology in developing countries. *Technology in Society Journal* 20(3): 371–375.

Wicklein, Robert C. 2004. Design criteria for sustainable development in appropriate technology. Paper presented in the PATT-14 Conference, Albuquerque, New Mexico, March 18–20, 2004.

Wieseltier, Leon. 2015. Among the disrupted. *New York Times*. http://www.nytimes.com/2015/01/18/books/review/among-the-disrupted.html?ref=books&_r=1. Accessed January 10, 2016.

Wolfe, Cary 2010. *What is Posthumanism?* Minneapolis: University of Minnesota Press.

Index

Notes
- The various friends and colleagues listed in the preface and acknowledgements are omitted from this list, unless they are also mentioned throughout the text
- Not only human proper names, but also non-human ones (such as Lana, Washoe, Koko…) are listed

D. Martinelli, *Arts and Humanities in Progress*, Numanities - Arts and Humanities in Progress 1, DOI 10.1007/978-3-319-45553-2

MIX
Papier aus verantwortungsvollen Quellen
Paper from responsible sources
FSC® C105338

If you have any concerns about our products,
you can contact us on
ProductSafety@springernature.com

In case Publisher is established outside the EU,
the EU authorized representative is:
Springer Nature Customer Service Center GmbH
Europaplatz 3, 69115 Heidelberg, Germany

Printed by Libri Plureos GmbH
in Hamburg, Germany